Dear Reader,

In March 2004 I was in a restaurant in London. A disclaimer proudly indicated that no genetically modified food was used as an ingredient in anything on the menu. Of course, to today's typical European that disclaimer is commonplace. It wouldn't be here in the United States, where the debate over the value and safely of GM foods has settled down quite a bit. But it was a shock to me! For although I concede happily the good sense of being careful in applying any scientific breakthrough—especially any scientific breakthrough I put in my mouth—everything I'd read up till then in the newspapers and science magazines had given me the impression that GM foods are perfectly safe and nutritious; indeed, that they could be of enormous benefit to farmers around the world. Suddenly, though, I found myself in the position of being reassured that the concerned owners of this restaurant were protecting me from the fruits of decades of scientific progress. How nice of them!

So what's the problem?! Fear? Politics? Economics?

If you did a Web search you'd probably have difficulty separating the ignorance and political agendas from the facts and the good intentions. The aim of this book, co-authored by one of the world's leading experts on genetically modified organisms and one of the most talented science writers on the planet, is to give you the facts behind all the hype and scare tactics, so you can decide for yourself whether you should support or block research on, and distribution of, genetically modified foods. After reading *Mendel in the Kitchen,* you will have a better understanding of the science and perhaps the promise of genetically modified foods.

Good reading and bon appetit!

Jeffrey Robbins
Senior Editor
The Joseph Henry Press

MENDEL
IN
THE KITCHEN

MENDEL IN THE KITCHEN

A SCIENTIST'S VIEW OF GENETICALLY MODIFIED FOODS

Nina V. Fedoroff and Nancy Marie Brown

Joseph Henry Press
Washington, D.C.

Joseph Henry Press • 500 Fifth Street, NW • Washington, DC 20001

The Joseph Henry Press, an imprint of the National Academies Press, was created with the goal of making books on science, technology, and health more widely available to professionals and the public. Joseph Henry was one of the founders of the National Academy of Sciences and a leader in early American science.

Any opinions, findings, conclusions, or recommendations expressed in this volume are those of the author and do not necessarily reflect the views of the National Academy of Sciences or its affiliated institutions.

Library of Congress Cataloging-in-Publication Data (to come)

Cover image:

Illustrations by Jeffery Mathison

CONTENTS

ILLUSTRATIONS

PREFACE

. . . making the yellow soil express its summer thought in bean leaves and blossoms rather than in wormwood and piper and millet grass, making the earth say beans instead of grass—this was my daily work.

—Henry David Thoreau (1854)

Our civilization rests on food: on our ability to make the earth say beans, to store those beans and fruits and seeds, and to share them. Other creatures might feed their young, but as adults each one fends for itself, spending much of the day doing it. By contrast we humans have learned to farm. Over the last few centuries, advances in science have allowed fewer and fewer farmers to feed more and more people, freeing the rest of us to make and sell each other houses, hats, and video games, to be scientists and writers and politicians, painters, teachers, doctors, spiritual leaders, and talk-show hosts. In some parts of the world, only one person in 200 grows plants or raises animals for food. The other 199 of us buy what we eat.

Whether we tote home sacks from a supermarket or dine out in a restaurant, most of us never give a thought to the growing, processing, packaging, and shipping of our food. Nor are we overly concerned about its safety. We rarely get sick from eating what we buy. We are surprisingly unaware of what it takes to create our bread and breakfast cereal, pasta and rice, those perfect fruits and vegetables, unblemished by insect bites or fungal spots. We do not know what makes our agriculture so efficient and our food so cheap. We cannot tell why it is nutritious or safe to eat. Free to live our lives with little thought for our food, we ignore the source of the gift, the source of our civilization.

Our civilization rests, in fact, on a history of tinkering with nature, on making the earth say beans, as Thoreau so eloquently said, instead of grass.

Thoreau's beans were not wild. The pod of a wild bean bursts when its seeds are ripe, flinging the beans far from the parent plant to find a new place to sprout. The bean pods we grow for food do not burst so they can no longer seed themselves. Neither can the wild grasses we have changed over the millennia into our staple food sources: rice, wheat, and corn.

To change a wild plant into a food plant requires changes in the plant's genes. To boost its yield, to make the earth say more beans, means changing the plant's genes as well. For thousands of years people have been picking and choosing plants, propagating plants with genetic changes—mutations—that made them better food plants. Thoreau, of course, would not have thought of either beans or grass in terms of genes. He published his influential book, *Walden*, in which he describes his efforts to make the earth say beans, in 1854. Gregor Mendel's experiments with peas, which would give rise to the new science of genetics, had not yet been done. Mendel didn't publish his work until 1866, and its significance wasn't grasped until more than 30 years after that. Not until the twentieth century did farmers begin to understand that their successes and failures had to do with genes. Yet well before Mendel explained how it worked, farmers were changing plants' genes.

At the end of the Stone Age, when most people still lived in small tribes hunting wild game and gathering wild plants, the world's human population was stable at eight to ten million. Then, when farming took hold as a way of life, the population began to grow. By the time of Christ, it had risen to between 100 and 300 million. When Columbus landed in the New World and the spread of food plants around the globe increased, the world's population was about 450 million. By the late 1700s, when the new science of chemistry entered agriculture, it had doubled to 900 million. A century later, when Mendel's experiments were rediscovered, the population of the world was more than one and a half billion.

In just the last hundred years—an instant in human history—the population doubled and redoubled. The number of people on earth reached three billion in 1950, then jumped to six billion in little more than a single human generation. Yet farmers kept pace through advances in plant breeding: plants' genes were modified in ways that capitalized on the nitrogen chemists had learned to pull out of the air.

From the 1960s to the 1990s the new crop varieties and expanding fertilizer use—the Green Revolution—continued to meet the world's food needs. In 1950 1.7 billion acres of farmland produced 692 million tons of grain. In 1992, with no real increase in the number of acres under cultivation, the world's farmers produced 1.9 billion tons of grain—a 170 percent increase. If India alone had rejected the high yielding varieties of the Green Revolution, another 100 million acres of farmland—an area the size of California—would have had to be plowed to produce the same amount of grain. That unfarmed land now protects the last of the tigers.

Yet as the twentieth century came to a close, plant breeders began running out of breeding room. The Green Revolution had largely run its course. The increases in the yields of corn, wheat, and rice began shrinking year by year. Earth's human population, on the other hand, was still growing fast. Eight to nine billion people are expected to populate the planet by 2050. Feeding them is a problem both daunting and complex. Crop yields must be increased simply to provide all people with the same amount of food available to us today—unless more land is brought into production, unless more wilderness is plowed.

As has happened before when famines were predicted, plant scientists searched for new ways to increase the earth's yield. This time their innovations aren't called plant breeding but "genetic engineering." The new crops are not known simply as crops—as were the ones created using earlier ways of modifying plant genes—but genetically engineered, genetically modified, genetically manipulated, transgenic, or genetically altered. Most often they are lumped under the acronyms

GM, for genetically modified, or GMO, for genetically modified organism.

GMOs have met with strong resistance. Before GMOs, people might have protested the use of synthetic fertilizers or pesticides in modern farming, but they were unconcerned about whatever it was that plant breeders had done to create high-yielding hybrid corn or brilliant red grapefruit or seedless watermelons or canola oil. Now, however, many people seem to agree with Britain's Prince Charles when he calls the new techniques of plant breeding "dangerous" and against God's plan. Why?

One reason for this resistance lies in the words themselves. Much human effort goes into changing our environment by building highways, houses, air conditioners, shopping malls, dams, or airplanes. Although individual projects might meet with resistance, few people protest this kind of engineering. Yet the notion that plants are being engineered caught people by surprise. It was rather disquieting. Plants are, after all, *natural*, aren't they? Might we not be messing with Mother Nature if we began to engineer plants?

Another reason is that most of us simply don't know what to make of the molecular techniques that allow scientists to change plant genes or add new ones. We don't really know, but we suspect it might be dangerous to transfer a gene from one species, such as a bacterium or a fish, to another, such as corn or tomato. Could it be morally wrong to violate the species barrier? What *is* a species barrier anyway?

What genetic engineering actually is and how it differs from earlier techniques of plant breeding is not understood by many outside the laboratory and breeding plot. Nor do most people understand the effects on the science of plant breeding of new interpretations of patent law and federal regulations concerning food safety and environmental protection. People have heard that scientists themselves oppose genetically modified foods—and a few do, although they are rarely those who know this new science well. Most people lack the time—and often the knowledge—to critically examine the scientific research cited in support of the opposing views of the technology. By writing this book we seek to answer the questions that most people—whether for or against the idea of genetically modified foods—often forget to ask.

We cannot turn the clock back. The human population is too large, and the earth too small, to sustain us in the ways our ancestors lived. Most of the land that is good for farming is already being farmed. Yet 80 million more humans are being added to the population each year. The challenge of the coming decades is to limit the destructive effects of agriculture even as we continue to coax ever more food from the earth. It is a task made less daunting by new knowledge and new methods—if we use them wisely.

AGAINST THE
WAYS OF NATURE

*One of the most remarkable features in our domes-
ticated races is that we see in them adaptation, not
indeed to the animal's or plant's own good, but to
man's use or fancy.*

—Charles Darwin (1859)

Golden Rice is a rice rich in beta carotene, the substance that gives carrots their color. Its creator, Swiss scientist Ingo Potrykus, wasn't trying for a colorful garden curiosity like the Iceberg blackberry, a paradoxically white blackberry created in the 1890s by plant breeder Luther Burbank (better known for creating the Idaho potato and Shasta daisy). Nor was Potrykus trying to complement the contemporary dinner-plate aesthetics of the All-Red potato, bred by Robert Lobitz in 1984, or the Graffiti cauliflower, introduced to American seed catalogs from Europe in 2002. The All-Red potato is prized by chefs for retaining its cranberry color even when cooked; it makes a striking potato salad. The Graffiti cauliflower's deep blue-purple hue is an "outrageous bid for attention, the vegetable equivalent of a Versace neckline," wrote one food critic. It was one of many new vegetables of 2002 "designed to stir excitement and desire in the hearts of consumers."

Potrykus was not even trying to improve upon an already colorful "sport," as breeders name those interesting mutant varieties that crop up from time to time; he was not like Richard Hensz, the Texas A&M University researcher who in the 1960s made pink grapefruit more red,

1

producing the bright and attractive Star Ruby and Rio Red. Potrykus was not trying to create the culinary delight of saffron rice on the stalk.

The parchment-yellow hue of Golden Rice—which, unlike the other colorful varieties mentioned, is not yet on the market—was merely a side effect. The color signals the rice's beta-carotene content. Beta carotene is a precursor of vitamin A. Yellow rice, Potrykus knew, could help the hundred million children who risk blindness and death from vitamin A deficiency each year in countries where rice is the staple food.

There are 22 species of rice. Two are cultivated. The most popular, *Oryza sativa*, is grown in more than a hundred countries, although most of it is eaten in Asia. There are more than 80,000 varieties of *Oryza sativa*. Some cultures prefer a sticky rice, some like it drier. Some want slender grains, some round. Some want an aromatic rice, some favor rice with a red or purple hue, which, if not more nutritious (as they believe), at least brings good luck (red, in Asia, is a lucky color). Rice is the staple food for nearly half the world's people. It provides more than half—sometimes almost all—of the calories eaten by people in Bangladesh, Cambodia, Laos, Myanmar, Thailand, Vietnam, and Indonesia. In those countries, infants being weaned eat very little else.

Yet a grain of white rice, polished the way most people like it, lacks crucial vitamins, among them vitamin A. The first vitamin to be named, vitamin A was discovered in 1915. The livers of animals and fish are a good source but so are green and yellow plants because the body can turn beta carotene into vitamin A. Early signs of vitamin A deficiency include difficulty seeing in dim light. The eyes become unusually dry. And yet by the time the eye's cells are damaged, many other bodily systems are in distress; someone suffering from vitamin A deficiency often succumbs to intestinal or respiratory disease. Each year as many as 230 million children are at risk of going blind from vitamin A deficiency. The World Health Organization estimates that more than a million children a year die for lack of vitamin A in their diets. To the Rockefeller Foundation, a philanthropy devoted to "applying science to benefit mankind," creating rice that contains vitamin A seemed a sensible way to help those children.

The idea first surfaced in 1984. At a meeting of rice breeders at the

International Rice Research Institute in the Philippines, Gary Toenniessen of the Rockefeller Foundation had asked, "What gene would you put into rice, if you could put in any gene at all?" The breeders were skeptical that they *could* put new genes into rice. Genetic engineering was quite a success in medicine—human insulin to treat diabetes was by then being produced almost exclusively in genetically engineered bacteria—but it was hardly routine in plants. The first practical technique had been published only the year before.

But Toenniessen, a microbiologist whose earlier work had centered on integrated pest management and biodegradable alternatives to pesticides, pushed the issue. "Finally, Peter Jennings said 'yellow endosperm,'" Toenniessen said, retelling the story 20 years later. "Peter was one of the creators of IR8, the rice that started the Green Revolution," he continued. "He was then working in Latin America. 'As long as I've been a rice breeder,' Peter said, 'over 20 years, I've been looking for a rice with yellow endosperm, because then it would produce vitamin A.'" If he had found such a rice, with beta carotene tinting the edible part of its kernels, he could have enriched commercial varieties through crossbreeding. But he had not found one. Rice, like all green plants, has beta carotene in its leaves, where the yellow pigment helps in photosynthesis; but none of it accumulates in the grain, unlike in maize (or corn), with its yellow kernels.

"The more I thought about that," Toenniessen said, "the more it struck me as doable. If Nature has figured out how to do it, then we can figure out how to do it. And Nature does it. Maize produces beta carotene in the endosperm with no deleterious impacts on other parts of the plant." And maize and rice, both grasses, are quite similar in many ways.

Toenniessen funded a study at Iowa State University to clone the maize genes needed for beta carotene to form in the kernels. He funded a project at the University of Liverpool to analyze rice endosperm to see if the precursors to beta carotene were there. Learning that rice grains do contain the genes for beta-carotene production, he funded work at the City University of New York to see if those silent genes could be turned on. And he organized a workshop, with "everyone in the world" who knew about beta-carotene biochemistry and genetics, to which he invited several "gene jockeys," as he calls them, who were willing to work on rice. At the workshop, Ingo Potrykus met Peter Beyer, an expert on beta carotene in daffodils. They put their heads together. In 1990 Toenniessen offered funding, which grew over 10 years to $2 million. In 1999 Potrykus and Beyer patented Golden Rice.

Those first grains were merely a "proof of concept," the inventors said. More tinkering would be needed before Golden Rice would be of any use in the effort to combat vitamin A deficiency. The prototype rice would be tested to see if it contained allergens, such as the one associated with "daffodil pickers' rash." The antibiotic-resistance marker gene the first grains carried would be taken out: it was of use in the laboratory, but not in the field or final product. The amount of beta carotene would be increased, perhaps by using a different promoter to express the genes. And the new genes would be bred into easily grown strains of rice, which would then be thoroughly field-tested. Still, it seemed feasible that Golden Rice could eventually provide 20 to 50 percent of a child's daily requirement of vitamin A, enough to make up for the typical deficiency and to prevent a child who ate it every day from going blind. According to one collaborator, "We are aiming the benefits of Golden Rice at the poorest of the poor who cannot get anything other than rice, green chilies, and salt, if at all."

Ingo Potrykus is typical of many modern plant scientists around the world. His story—the story of Golden Rice—could belong to almost any one of them. He was not a plant breeder like Luther Burbank, working the soil, although greenhouses and growlights were a vital part of his lab. He was a geneticist whose research subject was not the fruit fly, the mouse, or the worm *C. elegans*; not bacteria or bacteriophages or molds, but plants.

Potrykus began his study of genes in the early 1970s, when he tried to turn a white petunia pink. He wanted to see if a gene could be taken from one plant and inserted into the cells of another, as was then being done routinely in bacteria. His results, in 1972, were baffling and never published. By then enough failures had accumulated that most scientists agreed plants could not exchange genes the way bacteria did. The only way plant genes would recombine, they concluded, was through pollination: plant sex. Not for several more years would studies of the plant pest *Agrobacterium tumefaciens* suggest a very different approach to gene transfer, one not so much like plant sex as *bacterial* sex.

Yet Potrykus persevered, even after the *Agrobacterium* method was announced in 1983. Twelve years after he had first tried to transfer genes between plant cells, he succeeded, this time using tobacco, which, like petunia, was a favorite plant for such studies. (It has since been eclipsed by a little weed in the mustard family, *Arabidopsis thaliana* or mouse-ear cress.) In 1984 Potrykus's lab, then in Basel, Switzerland, published "an ironclad demonstration" and "the first incontrovertible evidence," according to Paul Lurquin, an American scientist who reviewed the work, that plant cells could take up and incorporate foreign genes directly. Potrykus's proof, coming within a year of the first report of gene transfer using *Agrobacterium*, finally answered the question raised in 1967: Could genes be introduced into plants? It was, said Lurquin, a "turning point announcing an era of applications rather than one of basic discovery."

Potrykus decided the application he would pursue was to help developing countries grow more food. Together with a colleague, he set up a new Institute of Plant Sciences in Zurich. Even in the early 1970s, he wrote, there had been "claims (from those working with tobacco and petunia) that the new techniques would contribute to food security in developing countries. Obviously, to contribute, one would have to work with important crop plants and not only talk about them." He began projects to increase the yields of wheat, corn, barley, and rye by making them more resistant to disease. Two Indian scientists, Karabi Datta and her husband, Swapan Datta, drawn to the new lab, persuaded Potrykus to add rice to his list of experimental crops and to shift his emphasis away from yield alone. Rice, Swapan Datta argued, is food for the poor. When improving it, nutrition should be considered, not just yield, which had been the main objective of the breeding that produced the Green Revolution varieties of the 1960s.

From 1990 to 1999, with funding from the Rockefeller Foundation, Potrykus and his colleagues worked to transfer into rice the genes that give daffodils their brilliant yellow color—the genes needed to make beta carotene. It was not a simple task. Until then only one or two new genes at a time had been added to a plant. To make beta carotene, four genes had to be added and their protein products had to interact. It proved more difficult than anyone had anticipated. Eventually even Peter Beyer gave up hope—although he didn't let Potrykus know. "This exemplifies the advantage of my ignorance and naivete," Potrykus wrote later. "With my simple engineering mind I was throughout optimistic and, therefore, carried the project through. I was naive enough to believe in its success." But it took nearly 10 years and the concentrated work of some 60 scientists from Germany, Switzerland, Poland, India, Japan, and China before Potrykus saw the first golden grains—at a symposium celebrating his mandatory retirement from the lab when he reached the official Swiss retirement age of 65.

The timing seemed perfect. In France and Ireland, protesters had vandalized seed stores and test plots of what had come to be called genetically modified organisms (GMOs) or genetically modified (GM) food. Austria, Luxembourg, and Norway had banned the planting of GM corn. Food stores in Britain had pronounced themselves GM-free, and fast-food outlets soon followed suit. Britain's Prince Charles had accused scientists of interfering with God's plan, warning of a "manmade disaster," "unforeseen consequences," and threats to biodiversity. In the U.S., where almost half of the soybeans and cotton and 25 percent of the corn planted in 1998 was genetically modified for insect or herbicide resistance, a lawyer, twelve religious leaders, and nine university scientists had brought suit against the U.S. Food and Drug Administration (FDA) for its "negligent oversight" and "unethical" handling of genetically modified foods.

But Golden Rice, Potrykus thought, was GM with a difference. Publicly funded, it would be given away free to subsistence farmers; multinational corporations would make no profit from the poor. The farmers could save the seeds from their crops, and those seeds would breed true the next season. The rice would provide better nutrition— but otherwise the plant was unchanged. It didn't require more fertilizer or pesticide. "To grow this rice does not require any additional input," Potrykus has said. "All a farmer needs to benefit from the technology is one seed." Because the trait could be bred into any of the thousands of varieties of *Oryza sativa*, it need not restrict the rice gene pool. In fact, it would be fairer to speak of golden *rices*, not one Golden Rice.

Golden Rice would change the international debate. With pride, Potrykus and Beyer applied for a patent, licensed it to Zeneca Agrochemicals, then sent a paper off to the prestigious British journal, *Nature*. The journal rejected it. Potrykus persevered. He garnered letters of support from famous European scientists. *Nature* declined even to send the paper out for review.

Word of Golden Rice reached Peter Raven, head of the Missouri Botanical Garden and organizer of the International Botanical Congress planned for St. Louis in August 1999. Raven invited Potrykus to speak at the meeting, held a press conference, and encouraged the American journal *Science* (equal in prestige to *Nature*) to read the manuscript. *Science* accepted it, printing it in January 2000 with a glowing commentary. It was a "tour de force." It "exemplifies the best that agricultural biotechnology has to offer," the reviewer said. By July Potrykus's face (looking puzzled but determined, his chin beard and brush mustache mostly white) had made the cover of *Time* magazine. "This rice could save a million kids a year," the banner headline read. The industry-sponsored Council for Biotechnology Information (whose slogan was "Good ideas are growing") produced a new ad— "Biotechnology researchers call it 'golden rice.' For the color. For the opportunity."—and began running it in such venues as the *New York Times* and the *New Yorker* magazine.

The backlash was fierce. Potrykus had created a "Frankenfood," a source of "genetic pollution." He had sold out to the multinational corporations (Zeneca Agrochemicals had by then merged with rival Novartis Agribusiness to become Syngenta, "the world's leading agribusiness.") Golden Rice was "fool's gold," "an intentional deception," a "gift horse"—even a Trojan horse—for the developing world. It was "a useless application" and "an exercise in how not to do science," wrote Mae-Wan Ho, a founder of the Institute of Science and Society and scientific advisor to a nongovernmental organization called Third World Network. Golden Rice was "a rip-off of the public trust," as well as a "major obstruction" to work that would "provide the real solutions to world hunger." It was "worse than telling them to eat cake."

As the vitriolic attacks continued, Potrykus began to wonder how he could have been so naive as to think that he could use science to solve a social problem. A press release from the Rural Advancement Foundation International, a "civil society organization" based in Canada, accused him of having "surrendered a decade of public funding to the commercial and PR interests of the biotech industry." In response, Potrykus plaintively wrote: "That I tried to use the funds for both competitive science and contribution to food security in developing countries was my personal and free decision. I could have used the same funds for studying why the hairs on the leaves of the small weed *Arabidopsis thaliana* are sometimes two- and sometimes three-forked." If he had done so, he mused, perhaps no one would have complained.

The term "Frankenfoods," with its power to call up the mad scientist Frankenstein and his unnatural monster, was first applied to genetically modified food by a Boston College English professor in a letter to the *New York Times* in 1992. Frankenfoods has not clarified the debate or contributed to, in Potrykus's words, "any discussion of a risk/benefit analysis." Just the opposite. "The Frankenstein myth," writes critic Jon Turney in *Frankenstein's Footsteps: Science, Genetics, and Popular Culture*, "hampers the necessary task of agreeing how to control the new technological powers now being developed in the laboratories.

"It invites an all-or-nothing response to a whole complex of developments, when we should be insisting on our right to choose some, and block others. When we do so, it should be for reasons which we can articulate more clearly than saying either that there are some things humans are not meant to know, or that we should not tamper with nature."

How did Luther Burbank create a white blackberry? How did Richard Hensz create the grapefruit called Rio Red? Few people know. The history of food plants is not a popular subject. Few people recognize the art and artifice in plant breeding—that each new variety is a product of the human mind and hand. It is hardly common knowledge that farming and science have been intertwined for 200 years, and that well before then, more than 10,000 years ago, the way humans procured their food became distinctly unnatural. Yet to think clearly about Golden Rice, and about genetically modified foods in general, we need to know these things. We need to know that people have been genetically modifying plants for many thousands of years. We need to understand the science behind these genetic changes, and the meanings people give to the terms they use to describe them.

We need to be able to compare, for instance, the risks and benefits of Golden Rice—which brings together genes from two plants belonging to two unrelated species, rice and daffodil—with those of triticale—a cross between two unrelated species, wheat and rye. Each was produced in the laboratory: Golden Rice by molecular techniques, triticale through chemical ones. Triticale, says a botany textbook, is the "first successful attempt to synthesize a new crop species from intergeneric hybridization"—meaning that it is the first artificial creation of a hybrid between two plants not only not in the same species, but not in the same genus. Yet triticale flour is available in the natural-foods section of the grocery store. Its label touts its nutritional goodness. Golden Rice, on the other hand, is condemned as Frankenfood and genetic pollution.

We have all eaten a red grapefruit. Who knows—or cares—that the most popular red variety, the Rio Red, was created by exposing grapefruit buds to thermal neutron radiation at Brookhaven National Laboratory in 1968? Nor do we recall that the original pink grapefruit was a mutant that appeared as a sport on the limb of one grapefruit tree in a Florida orchard in 1907 and has been cloned countless times since then to make all the pink grapefruits in the world.

Proponents of genetically modified foods often argue that the molecular techniques lumped under the term genetic engineering or genetic modification are simply an extension of conventional plant breeding. According to Klaus Ammann, curator of the botanical garden at the University of Bern in Switzerland, "The most significant changes in grains and advances in knowledge about crop genomes occurred many years ago. When we eat wheat, we consume varieties mutated by nuclear radiation. It is not known what happened with the genomes, but we have been eating this wheat for decades, without any type of problem. Today, with more extensive knowledge and new applications of the technologies resulting from genetic engineering, we are faced with a new system where control is greater, more precise, and less risky than that of the old systems."

Opponents counter that genetically modified food is unlike anything that has been produced before. Dennis J. Kucinich, a U.S. representative from Ohio, argued in a recent video production that genetic engineering "is not the same as conventional growth of food. It's a manmade process. It has nothing to do with the ways of nature. It's very violent." Which view will seem right to you depends on what you consider conventional, and on how you define the ways of nature.

By 1860, for instance, German botanist Julius von Sachs was growing crops hydroponically—in water, not soil. Though his technique was "much disputed at first," he successfully grew garden beans, field beans, corn, and buckwheat in water, "in a sunny window or in a suitable greenhouse," produced fertile seeds, and planted and grew the offspring. He set a corn kernel in a box "filled with well-washed damp sawdust" until it germinated and sprouted. "After the seedling has been carefully taken out and washed, it is fastened into a perforated cork . . . so that only the root dips into the water." To the water he added various chemicals containing nitrogen, iron, potassium, calcium, magnesium, phosphorus, and sulfur. It was the beginning, he wrote, of a "complete revolution" in agriculture. "We are now able to rear plants artificially," he reflected in 1887. From "inconspicuous and often scarcely ponderable quantities of vegetable substance," scientists could produce quantities of food "as large as we choose."

According to Harvard biologist John Torrey, Sachs's work was the beginning of plant biotechnology. The next step was straightforward. Once plants had been freed of soil, he writes, "it was natural to ask whether it was possible to grow plant parts independent of each other." In fact, in 1902, German botanist Gottlieb Haberlandt theorized that each and every plant cell contained all the necessary instructions to grow a complete plant. It should be possible to grow a plant not only from a seed but from a bit of leaf or a root tip. Plant scientists tried growing the tips of tomato and pea roots in all sorts of solutions, including yeast extract and beef broth. Eventually they found that the roots required tiny amounts of certain vitamins, in addition to sugar (ordinarily made by the plants themselves through photosynthesis) and the minerals Sachs and later experimenters had identified. These would keep a root tip alive and growing—but only as a root.

To get a shoot and leaves required plant hormones. Charles Darwin and his son Francis, in 1881, had noted an "influence" that caused seedlings to bend toward the light. In 1926 the Dutch plant physiologist Frits W. Went proved it to be one of a class of plant hormones he named auxins (from the Greek, "to increase"), because he found that the hormone controlled how cells lengthen. French scientist Roger Gauthieret discovered that auxins could keep plant tissues growing indefinitely. In the mid-1930s he took chunks of carrot and cultured them in a solution of sugar, minerals, and vitamins. When he added auxins, an unorganized mass of cells grew at the cut ends of the chunk. He called this mass "callus" tissue because it resembled the calluses that grew around wounds on tree trunks. Plucked off and set in a separate petri dish with the standard nutrients, the callus continued growing—in this lumpy, undifferentiated state—as long as it was supplied with auxin. One of Gautheret's carrot cultures survived for 40 years. But it was still just a mass of cells, not a plant. Scientists kept searching for the substance that would turn a callus into roots and shoots, leaves and fruit.

In the 1950s F. C. Steward at Cornell University was growing tiny bits of carrot root in a solution containing coconut milk. The mass of carrot cells, just like Gauthieret's, grew and divided. When Steward swirled the flask, some cells broke free from the mass. These grew and divided and occasionally formed roots. If he transferred them to a solid agar gel, some of them developed shoots. If set in earth the roots and shoots grew into plants that flowered and set seed. The hormone involved in this case was identified in 1954 by Carlos O. Miller, a postdoctoral researcher at the University of Wisconsin, when, in desperation after trying everything else, he squirted drops from an old bottle marked "Herring Sperm DNA" onto his culture medium, and noticed that it caused his tobacco cells to divide.

Miller and his mentor, Folke Skoog, had, like Steward, originally tried coconut milk, but they could not purify the growth-promoting factor from it. Knowing that a similar factor existed in herring sperm DNA (but only in *old* herring sperm—new samples had no effect on the tobacco cells) gave Skoog and Miller the hint they needed to isolate the hormone. They named it kinetin and called the family the cytokinins, because of their role in cytokinesis, or cell division. Cytokinins are derivatives of nucleotides, the building blocks of DNA; the herring sperm DNA worked because it was so old that much of it had broken down into its basic parts. Skoog and Miller showed how the two plant hormones, the auxins and the cytokinins, governed the transformation of a callus into a plant. When the two hormones were in balance, the clump of cells remained a callus. With less cytokinin, roots grew; with more cytokinin, buds and shoots formed.

These new insights into how plant cells grow and differentiate were brought into farming by French botanist Georges Morel. In the late 1950s Morel was trying to rid potatoes of viruses that stunted the plants and caused deformed leaves. Even in heavily infected plants, he found, the very tip of the growing shoot, the meristem, was free of disease. Morel snipped off the meristem (eventually other researchers would show that a sliver thinner than a fingernail was all that was needed) and cultured it. The cells multiplied and, when he let them grow on a solid medium with the right balance of hormones, they turned into thousands of genetically identical plants—clones of the original, but now virus-free. The technique came to be known as tissue culture cloning.

One of Morel's graduate students, Walter Bertsch, "happened to be an orchid fancier and also happened to be dating a girl who worked at a famous French orchid company," writes Susan Orlean in her book *The Orchid Thief.* "This is how orchids came to be the first ornamental plant to be cloned on a large scale." Orchids had been prized not only for their rare beauty, but also for the skill it took to grow them. "With cloning, they could be almost as common as daisies." The orchids now routinely sold in grocery stores and discount outlets are all clones produced through tissue culture.

Scientists in Japan learned how to use these techniques to "rescue" rice embryos from hybrid seeds that ordinarily would not germinate.

Soon they were making wide crosses between rice and wild relatives with which the cultivated varieties normally did not interbreed. Another type of tissue culture, the one that Peter Raven of the Missouri Botanical Garden has called "the most elegant," is protoplast culture. In 1960 English scientist E. C. Cocking used enzymes from bacteria and fungi to strip the sturdy wall off a cell from a tomato's root tip. In 1968 I. Takebe of Japan, using a tobacco cell, simplified the procedure. Seen through a microscope, these naked protoplasts (named for the Greek for "first," or "fundamental") are round green globes, not the squarish boxes of cells inside a plant. Their now-flexible membranes can be disrupted briefly to slip in a molecule, a virus, or even part of a cell, like the DNA-rich nucleus. When cultured under the right conditions, protoplasts regenerate their cell walls and divide, eventually growing into plants, some with new genetic characteristics.

To make new hybrids, scientists began trying to fuse cells of two different plants. In 1978 Georg Melchers of Germany succeeded in making a Tomoffel or Karmate, from the German names of tomato and potato. The half-tomato, half-potato protoplasts grew into plants that flowered but didn't set seed. The experiment would not have been surprising to Luther Burbank, who noted in 1893 that "tomatoes may be grown from seed pollinated from potato pollen only" and claimed in his catalogs to have thousands of tomato-potato hybrids growing in his plots. Nor has the effort to make a Tomoffel been abandoned. Laboratory specimens were grown to maturity in 1995 and the fertile offspring backcrossed to the parent plants. Crosses between a tomato and a wild eggplant have also been successful.

How does protoplast fusion work? T. Kinoshita and K. Mori of Hokkaido University in Japan reported in 2002 on their efforts to cross cultivated rice, *Oryza sativa*, with its close relatives in the genus *Oryza* in order to bring in "important genes for sustainable agriculture, such as resistance to brown planthopper, bacterial leaf blight, and grassy stunt virus." First they treated the protoplasts they had taken from rice seeds with certain chemicals, then chilled them at 4°C, a temperature just above freezing. This kept the cells from dividing. Then they positioned a wild protoplast next to a cultivated one in a test tube and applied an electrical current. The membranes of the different proto-

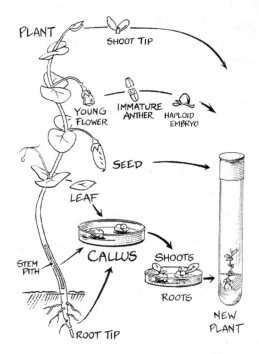

How to regenerate a plant

plasts fused. They exposed the fused cells to gamma radiation, after which they followed standard tissue culture protocols to grow some of the fused cells into mature plants.

Plants have also been grown from immature anthers—the part of a flower that makes pollen. A group of scientists from Hangzhou, China, reported in 2002 that anther culture was an effective way to improve rice cultivars. Their method involved clipping off the flowering tip, or boot, of the rice plant in the latter stages of pollen development. They wrapped the rice boots in aluminum foil and chilled them at 7ºC for five days, then disinfected them with successive treatments of alcohol, Chlorox, and distilled water. They placed 50 anthers in each of 10 test tubes containing some jelled growth medium, and incubated them at room temperature in the dark for 30 days. Then they irradiated the test tubes with gamma rays. With the gamma-ray treatment they got almost 500 green plants for every 100 anthers, 300 times the number that grew without irradiation. The bits of callus tissue that began to grow were then transferred into larger tubes and returned to the warm room, where they were grown under constant light for 40 days. The green plantlets that formed were fed nutrients and allowed to grow until their shoot tips reached the tops of the tubes, then their tubes were opened and they were moved into the greenhouse. A week later, they were transplanted into the field.

An interesting puzzle about plants grown from cultured tissues is that they are not all the same. Depending on the type of plant, its age, where the tissue was taken from (root tip, leaf, anther, elsewhere), the balance of hormones and nutrients in the test tube, and the length of time in culture, various types of mutations commonly arise even without the use of radiation or mutagenic chemicals. These mutations, which turn out to be of all the different kinds known to occur, are lumped under the term somaclonal variation (*soma* is Greek for "body" or "tissue"; a clone is a copy). As one scientist notes, "Uniformity among the plantlets regenerated from callus tissue is considered to be the exception rather than the rule."

Most of these mutations harm the plant. Plants with such harmful mutations are discarded. But occasionally a good mutation appears, and some have given rise to commercial crops. Clearfield corn, for instance, sold in the U.S. since the early 1990s, is a variety of maize resistant to Patriot, an imidazolinone herbicide. Another variety of maize to come from somaclonal variation tolerates glyphosate, the generic name of Monsanto's herbicide Roundup. Farmers who plant herbicide-tolerant varieties can spray their fields to kill weeds after the corn has sprouted without killing off the young crop. This practice is popularly thought to be limited to Roundup Ready crops that are genetically modified by molecular techniques. It isn't.

How natural are these plants? Are tissue culture-derived mutants more natural than plants to which an herbicide-resistance gene has been added? The answer, again, depends on the meaning of the word "natural." What is quite clear is that all of the techniques of tissue culture just described are considered conventional. Mutant plants arising by somaclonal variation in tissue culture are not given the GM label.

The use of radiation and chemicals is not limited to encouraging plants to grow from pollen cells in test tubes. At about the same time that Gottlieb Haberlandt first suggested that every plant cell could become a plant, Hugo de Vries, author of the influential book *The Mutation Theory*, proposed that X-rays could be used to mutate plants for study. In 1927 Hermann Joseph Muller proved that irradiating plants did cause mutations, and in 1928 Louis Stadler published the first paper on the effects of X-rays and radium on plant seeds. Soon other types of radiation, including gamma rays, fast neutrons, and thermal neutrons, were being used to cause mutations in plants. Chemicals were also used to induce new mutations, as well as to allow hybridization of varieties that would not naturally crossbreed.

The chemical colchicine, for instance, was isolated from the autumn crocus or meadow saffron in 1936 and found to be a pesticide. The news that it also doubled a plant's chromosomes "became the start of a real colchicine craze," according to one plant breeder. "Within a few years the chromosome doubling technique had been applied to at least 50 plant species." Applied to a sterile hybrid of rye and durum wheat in the 1950s, colchicine allowed F. G. O'Mara of Iowa State University to create the new, fertile grain triticale, which by the mid-1980s was being grown on more than two million acres in the Soviet Union, Europe (particularly France), the United States, Canada, and South America. The label for triticale flour from Bob's Red Mill Natural Foods in Milwaukie, Oregon, reads: "Triticale is a hybrid grain—a cross between wheat and rye. It averages 28 percent higher protein than wheat and contains all the essential amino acids, thus making it a more complete protein than the parent grains. It has an interesting nutty flavor and is high in fiber. . . . Since Triticale Flour has inherited the best qualities of its parental grains—wheat and rye—and comes with a delicious flavor all its own, it needs to be discovered. Right now . . . by you . . ." The label makes no mention of the fact that wheat and rye could not naturally hybridize.

Colchicine is also used to make fruits seedless. A favorite fruit produced this way is the seedless watermelon. Colchicine interrupts cell division, so that the cell makes extra copies of its chromosomes, and thus of the genes on those chromosomes, but fails to divide as it normally should. As a result, the seeds don't develop and the watermelon is much easier to eat.

Other notable successes of mutation breeding include the most popular variety of wheat used for making pasta in Italy. The durum wheat called Creso is a cross between two earlier types, each created by exposing seeds to neutrons or X-rays. It has "high and stable yield capacity, good adaptability and grain quality." A third of the area planted in durum wheat in Italy in the 1990s was planted in Creso.

Golden Promise barley, created in 1956 by exposing an earlier variety to gamma rays, "became a standard for malt quality and is still grown today under contract for specialist distillers" of fine beers, including some organic beers. B. P. Forster of the Scottish Crop Research Institute notes that Golden Promise was on the recommended list of cereals for England in 1967 and for Scotland in 1968. "It dominated Scottish barley acreage," Forster writes, "from the 1970s to the mid-1980s," even though it was susceptible to powdery mildew and needed to be sprayed with fungicides. Writes Forster, "The introduction of Golden Promise to Scotland revolutionised cereal growing to the benefit of the farmer, maltster, brewer, and distiller (fungicide manufacturers also profited), and enabled Scotland to come to the fore as a producer of malting barley."

Although the mildew problem eventually reduced its popularity, Golden Promise became the breeding stock for 15 other successful cultivars and might yet make a comeback in other countries: Forster recently discovered that it is highly salt-tolerant.

Some California rice is also a product of mutation breeding. A popular variety called Calrose 76, released in 1976, was created by treating seeds with gamma rays from a cobalt-60 source. In the early 1980s the high yielding Calrose 76 was earning California rice growers an extra $20 million per year.

And, as Swiss botanist Klaus Ammann hinted, of the hundreds of varieties of bread wheat grown around the world almost 200 were created using X-rays, gamma rays, neutrons, or various chemicals to cause mutations. The latest variety, a hard red winter wheat called Above, was mutated so that it would tolerate an herbicide produced by the BASF corporation. Just like Clearfield corn, Above wheat can be sprayed with herbicide and will not die, letting farmers weed their fields chemically after the crop sprouts. Yet, although the end result is the same as the Roundup Ready crops sold by Monsanto, Above is not considered genetically modified.

Several new crops, including a tasty little lettuce called Icecube and a number of varieties of barley used for making fine beers, were created by exposing seeds or plant parts to the chemical ethylmethyl-sulphonate. This chemical is known to cause genetic mutations in people as well, including some that cause cancer. Plant breeders who use it must wear coveralls and breathe through a respirator.

And yet, other than general safety recommendations for laboratory personnel, there is little government oversight of this type of plant breeding, whether for food crops or for ornamentals. When Susan Orlean, chronicling Florida's orchid industry, visited one backyard breeder, she reported: "One bench was stacked high with small plastic pots. The plants inside them were withered and droopy, a jumbled mess. Mike nodded toward it and said, 'Failed *Antherium* project. A species called Elaine. It was created by irradiation. We took the germinating material and radiated it. We hoped to get some interesting mutations, but it didn't work out that way.' I asked him what they were going to do with all the loser Elaines. 'Take all ten thousand of them and toss them in the Dumpster,' he said." Less than 1 percent of all observed mutations created through mutation breeding are of value. "Many induced mutations have direct or indirect adverse effects," one researcher noted. There is no official, recommended procedure for destroying the so-called "losers."

This lackadaisical oversight has not caused environmental problems even though, according to Henry Miller, a fellow at Stanford University's Hoover Institution, "Literally millions of genetically altered, but not gene-spliced, plants are field tested each year without governmental oversight or strictures: the average plant breeder of corn, soybean, wheat, or potato, for example, may put into the field 50,000 discrete, new genetic variants per year, many or all of which may be the product of 'wide crosses,' hybridization in which genetic material (including that from weedy or poisonous plants) has been transferred across natural breeding barriers." Fifty thousand might be a low estimate. In 1959 W. C. Gregory grew a million M2 peanut plants (M2 meaning that they were the second generation after mutagenic treatment of seeds) in North Carolina. Among the million was one that became a commercial cultivar, registered in 1964 as N.C. 4X. And at the Carlsberg Research Laboratory in Copenhagen, as of 1995, Dieter von Wettstein and his colleagues had screened 18.5 million chemically mutated barley plants, looking for one that did not cause chill haze in beer.

Nor, as Ammann notes, has human health suffered from eating bread made from mutated wheat or drinking beer made with mutated barley. "Nowadays, for a number of crops, one even has lost track of the newly induced mutant cultivars," says A. M. van Harten, who has taught mutation breeding for 25 years in the Netherlands. "Breeders often do not care to distinguish anymore between spontaneous and induced mutants. Part of the reason for this is that the public has become aware of—supposed or real—risks of what is commonly called 'biotechnology.' As a consequence, breeders may prefer not to mention anymore that their cultivars arose from mutagenic treatments, as this may negatively influence their customers." In other cases, he notes, the breeders might not even know that the building blocks, or parent varieties they are using in their breeding programs were mutation-bred.

None of the crops produced through the use of radiation or chemicals is labeled "mutation bred." Many, like triticale, are called, instead, natural foods. If a law were passed requiring certain crops to be labeled GM or GMO, these crops would not be included. These procedures are all considered modern standard breeding techniques. "The term itself conveys a sense of slow-paced, gentle manipulations that are in some way natural," notes David Saul, a biologist at the University of Auckland, New Zealand. "But have no illusions about these procedures," he continues, "they are often genetically brutal and highly unpredictable." Some of them have been in use for 80 years or more, he notes. They "are the tools that have wrought the fruit and vegetables found in your typical organic food store."

According to the Mutant Variety Database (available on the

Internet), more than 2,000 crop varieties have been created using chemical or radiation mutation; of these, half were released since 1985. These crops, including varieties of wheat, rice, oats, barley, grapefruit, lettuce, and beans, are being grown in gardens and orchards and farm plots around the world. Are these crops any less genetically modified—or any more natural—than Golden Rice?

When W. Navarro Alvarez was trying to breed a salt-tolerant rice in Costa Rica using anther culture, he irradiated seeds of seven commercial rice varieties with gamma rays. He reported in 2002 that "all irradiated seeds were sown in the field and grown" for two generations. Out of 10,000 second generation plants growing on saline soil he found 64 mutants that were salt tolerant.

Yet in order to continue developing Golden Rice, scientists must work inside a Biosafety Level Four greenhouse, the same level of biosafety required of those who work with the deadly Ebola virus or anthrax. At the International Rice Research Institute in the Philippines, Karabi and Swapan Datta, the Indian couple who first urged Ingo Potrykus to work on rice, are breeding the golden color—and the genes coding for beta carotene—into 10 rice varieties currently grown in Bangladesh, Vietnam, India, the Philippines, and Mozambique. They chose "popular and successful plants within particular environments, plants with which we are totally familiar," explains Swapan Datta. One, for instance, is a rice called BR29 from Bangladesh. "It has good cooking quality and moderate disease and pest resistance, and it is well and truly adapted to its environment. The farmers are happy with it, the market is happy with it."

If we condemn Golden Rice because it was created by a "manmade" process, as Congressman Kucinich says, that "has nothing to do with the ways of nature," what do we do with Creso wheat or triticale or the Rio Red grapefruit or Alvarez's salt-tolerant rice?

What do we do with Luther Burbank and his Iceberg blackberry? Burbank did not use hydroponics or tissue culture or mutation breeding or genetic engineering to create his new crops, yet they are just as manmade. As Liberty Hyde Bailey, dean of agriculture at Cornell University, wrote in 1901:

His new plants are the results of downright, earnest, long-continued effort. . . . Before the flower opens he cuts off the petals. Thus the bees are not attracted, and they have no foothold. Then he applies the pollen with a free hand. . . . Mr. Burbank gets unusual hybrids because he crosses great numbers of flowers and uses much pollen. He is skillful in the technique. He also dares. He has no traditional limitations. He knows no cross that he may not attempt. He has not studied the books. He has not been taught. Therefore he is free. The professor of horticulture would consider it beyond all bounds of academic and botanical propriety to try to cross an apple on a blackberry; but Luther Burbank would make the attempt as naturally as he would dig a new lily from the fields.

Is this amount of interference, this amount of manipulation, still "natural"?

2 THE WILD AND THE SOWN

There is always enough seed escaping the harvest to maintain the stand. Domestication begins when the people plant that which has been harvested. Then there are two populations, the wild and the sown.

—Jack R. Harlan (1999)

In 1786 Antoine Augustin Parmentier received permission from King Louis XVI of France to grow potatoes on 50 sandy acres outside Paris. Potatoes, at the time, were not popular. French fries had not yet been invented. Rather, potatoes were said to cause leprosy, cholera, scrofula, rickets, and tuberculosis. Being roots, they upset the body's balance. They corrupted the blood; they caused flatulence. Relatives of mandrake, henbane, nightshade, and belladonna, they were undoubtedly poisonous—or aphrodisiacs, as the English had thought since Shakespeare's day. They were unfit for any but beasts and, being pig fodder, would lower the standard of living if humans ate them. Besides, no good farmer would grow potatoes. Not only weren't they fit to eat, they ruined the soil.

Parmentier disagreed. He persuaded Queen Marie-Antoinette to put potato flowers in her posies to rehabilitate the image of the poor plant. He asked for a sandy wasteland as his trial garden to show how well potatoes could grow in land useless for other crops. He then asked the king to post a royal guard. The soldiers stood guard only during the day. Each night, Parmentier was pleased to learn, peasants crept in while the guard was gone and stole potatoes—to plant, Parmentier

23

hoped, or at least to eat. Parmentier, a pharmacist by trade, had argued since 1770 that potatoes could replace bread when the wheat harvests failed, as those harvests quite often did in eighteenth-century Europe. But despite the success of his ruse—pretending his heavily guarded potatoes were only for royalty, not common folk—his idea caught on too late. Soon after his Paris experiment a bad harvest sparked the bread riots of the French Revolution, occasion for the famous line attributed to Marie-Antoinette, "Let them eat cake."

Potatoes were not new to Europe in the late 1700s. The Spanish had brought them from Peru two centuries before. Nor are they the only food now considered central to our lifestyle to be greeted at first with suspicion. A Jesuit priest wrote in 1590, "The main benefit of this cacao is a beverage which they make called Chocolate, which is a crazy thing valued in that country. It disgusts those who are not used to it, for it has a foam on top, or a scum-like bubbling." Likewise, in 1674 a group of English women described coffee, in a petition to government to ban coffee shops, as "base, black, thick, nasty, bitter, stinking, nauseous puddle water."

Just as they were chary of trying new foods, Europeans clung to the old ones when they emigrated to America. The tomato, a New World staple, was "treated with derision" and fed to the pigs as late as the nineteenth century. Instead, the settlers brought the foods they knew from their homelands: apples, oranges, figs, sugar, wine grapes, wheat, oats, barley, rice, cabbages, watermelon, and cantaloupe. They also imported honeybees, not found in America, to pollinate their new crops. To sustain their imported cattle, they imported grass seed, including English clover (which when it was introduced into England in the 1500s had been called French clover, the best varieties having come from France). Soon these imported grasses had taken over so completely—and often merely by the expanded grazing of cattle, who carried the seeds in their dung—that "within a generation of settlement," writes one historian, "many were believed to be native."

Nowadays, notes Edgar Anderson in his book *Plants, Man, and Life*, "Few Americans realize how completely our American meadow plants came along with us from the Old World. In our June meadows, timothy, redtop, and bluegrass, Old World grasses all three, are starred with Old World daisies, yarrow, buttercup, and hawkweeds. The clovers too, alsike and red and Dutch, all came from the Old World. Only the black-eyed Susans are indigenous. An informed botanist viewing such a June meadow may sometimes find it hard to point out a single species of plant which grew here in pre-Columbian times."

Even before the settlers came with their Old World fruit trees, grains, and grasses, humans had been remaking the American landscape. The Wampanoag of Massachusetts burned the forest underbrush each year to make it easier to find and follow game. In California, according to Florence Shipak of the University of Wisconsin-Parkside, the Kumeyaay harvested the grains of wild grasses, then burned the fields; after the autumn rains, they broadcast some of the harvested grain, along with seeds of green, leafy plants and other vegetables that would ripen at the same time. Other wild foods and medicinal plants they divided and transplanted to similar ecological niches, expanding their range. They planted oaks and desert palms, mesquite, wild plums, and pines whose nuts were edible. To destroy insects and diseases, they burned stands of agave, yucca, and other plants every five to ten years. These ways of domesticating whole landscapes—called firestick farming, hobby farming, or proto-farming—anthropologists now believe were practiced by humans for thousands of years before we domesticated individual plants. Rather than being a revolution as momentous as the discovery of fire, agriculture was the evolutionary next step in our everyday efforts to ensure our food supply.

Some theories say the step was taken for religious reasons: the best seeds were saved and planted to propitiate the gods. Others argue that humans were pushed over the edge by crowding or climate change or the loss of their major prey, or by all of these factors combined. It's no coincidence, they argue, that the earliest conclusive proof of true agriculture—a few grains of domesticated wheat found in the Near East—dates to the end of the last Ice Age. The glaciers were melting. Sea levels rose, affecting habitats around the globe. The Bering land bridge by which humans had crossed into North America disappeared. More than 70 percent of the large game animals in North and South America went extinct, including the mastodon, mammoth, ground sloth, peccary, several camels, the horse, and a guinea pig the size of a hippo.

Their disappearance was due to overhunting or to the change in climate, or to a combination of the two. In any case, it is likely that only those humans who already knew how to supplement their hunting with hobby farming could thrive. Such people in South America had already domesticated beans, peppers, and squash, and were on their way to adding tomatoes, corn, and potatoes. In China they were growing rice in paddies. In the Near East they found the wild grasses—especially wild wheat —particularly amenable to domestication because of their dense stands and large seeds.

The city of Jericho is far from the natural range of wild wheat. In prehistoric times its six acres, on the West Bank of the Jordan River, sheltered a population of at least 300—an extraordinary concentration of people for those days. A single family in a hunting and gathering culture of that period required a range of three square miles. Jericho, moreover, was built of stone and had an elaborate system of ditches and walls to divert floodwaters, as well as a 9-foot-high circular tower whose purpose we can only guess at.

By the time the tower was built 10,000 years ago, wheat had already gone through several changes at the hands of humans. The change most convincing to an archeologist is in how the spikelet holding the seed or grain attaches to the stalk. In wild wheat the spikelets shatter easily so that a light breeze can scatter the ripe seeds onto the ground. The end of a wild spikelet is smooth. In domesticated wheat the seed spike is not so brittle. The spikelets remain fixed to a miniature stem or "rachis" so strongly that no breeze can break them free. "The seeds 'wait' for the harvester," notes botanist Daniel Zohary. An archeologist can identify a grain that has been harvested and threshed by the rough scar on its end.

This change from a brittle to a sturdy rachis was caused by the mutation of one gene. To wild wheat such a mutation would be deadly: Its ripe seeds would dangle on the stalk, with no chance of being buried in soil, germinating, and continuing the species. But domestication in many ways inverts the rules of nature. The process has been likened to making plants into "wards" of people. Jared Diamond, a professor at

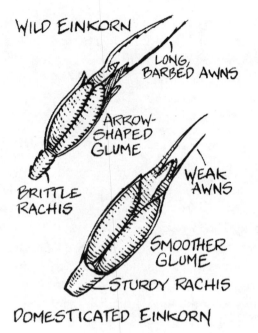

How to tell wild wheat from domesticated wheat

the University of California, Los Angeles, puts it this way in his book *Guns, Germs, and Steel*: "Human farmers reversed the direction of natural selection by 180 degrees: the formerly successful gene suddenly became lethal, and the lethal mutant became successful."

A second mutation that turned wild wheat into domesticated wheat was also lethal. The grains of primitive kinds of wheat are encased in tough hulls. What falls to the ground is not a smooth, rounded, naked kernel but the spikelet, "armed with thorn cells and rough barbs, all pointing upward," as botanist Jack R. Harlan describes them. Arrow-shaped, these spikelets "work their way into the soil, often aided by rough awns that vibrate in the wind." Once buried, the bristles have another function: the seed cannot germinate until the hull is broken down. Depending on the weather they are exposed to, seeds from the same summer will be ready to sprout at different times, sometimes years apart. This dormancy is the species' guarantee of survival. "If all the seeds sprouted with the first rains and there followed a long dry period before the next rains, the species would then become extinct. Some dormancy is required to build up a seed bank in the soil," writes Harlan in his book *The Living Fields*. Yet not only do the spikelets take some effort to thresh and winnow into grindable grains, the seeds' dormancy is a nuisance to the farmer, who wants all his seeds to sprout at the same time. Again, a mutation in a single gene made the difference between hulled wheat and naked wheat.

A third change farming made to wheat was in the size, shape, and make-up of the grains: they became larger, plumper, and better for making bread or pasta. The changes here were not due to the mutation of a gene, but to the addition of an entire genome, the full set of chromosomes of another plant. Einkorn wheat, the least domesticated variety, is diploid (from the Greek *diplo*, which means twofold): each plant has two sets of chromosomes. Einkorn inherits one set of 7 chromosomes from the male parent and one set of 7 from the female parent, for a total of 14. Because a human is also diploid, inheriting 23 chromosomes from father and 23 from mother for a total of 46, we tend to think of this situation—two sets of chromosomes making up one genome per organism—as normal. But botanists and plant breeders know it is not always so.

Emmer wheat, which along with einkorn was found buried at Jericho, is tetraploid. Each emmer plant has four sets of 7 chromosomes, for a total genome of 28. The relationship between the diploid einkorn and the tetraploid emmer was worked out early in the twentieth century by the Japanese scientist T. Sakamura. Einkorn's genome of 14 chromosomes, in two sets of 7, Sakamura called AA. Emmer wheat had all of these AA chromosomes, but it also had a pair of completely different chromosomes, which he labeled BB. These chromosomes, botanists believe, came from a weed common in wheat fields, most likely a kind of goat grass (*Aegilops speltoides*). At some point *Triticum urartu*—which is very similar to modern einkorn wheat—and this weed had hybridized, even though they were different species not even grouped together in the same genus. Because the resulting hybrid, emmer wheat, had more chromosomes than either of its parent types, it could not breed with either one: it was an entirely new species. Emmer wheat is known by the species name *Triticum turgidum*.

Emmer wheat was quite popular in the ancient world—the Romans preferred it to barley—and its sticky flour is still a favorite for making pasta. Durum wheat is a modern variety of emmer. But the major wheat of the world, bread wheat, is yet another species, *Triticum aestivum*. It is a hybrid of emmer and "a bristle-headed little weed of the Near East," as botanist Edgar Anderson has described it: *Aegilops squarrosa*, another kind of goat grass. It is thus a hexaploid, containing six sets of chromosomes, and is labeled AABBDD. It contains all the chromosomes of emmer wheat (AABB) and all the chromosomes of the weed, *Aegilops squarrosa* (DD), for a total of 42 chromosomes—three complete genomes.

The most ancient variety of bread wheat is spelt. Jack Harlan, who made a career of studying crop evolution, believed that the original cross between emmer wheat and goat grass to produce spelt happened near the southern end of the Caspian Sea shortly after emmer wheat itself was domesticated. This cross made wheat both drought-resistant and cold-tolerant, traits that permit it to be grown in North America, far from its origins. It also increased the quantity of gluten proteins in the wheat—these are the proteins that give bread dough its elasticity and allow bread containing yeast to rise.

But spelt is a hulled wheat. Its grains have to be heated, or parched, to free the kernel from the hull for grinding, and this destroys the structure of the gluten proteins. To arrive at the modern varieties of bread wheats—and make possible the invention of soft, raised bread—required two gene mutations (one to an A-chromosome gene and one to a D-chromosome gene). These mutations allowed for naked grains, giving harvesters a smooth, rounded kernel for easy grinding into a flour high in gluten proteins.

When scientists suggested that bread wheat got its goodness from two nuisance weeds, they turned the whole history of wheat on its head. That history had been written by taxonomists, scientists who study the shape and structure and habits of a plant and assign it to a genus (of which the plural is "genera") and species. As Edgar Anderson, writing in the 1960s, remembers, "We had set out to study the origin of wheat, and the taxonomists had told us we were studying *Triticum*, and there were such and such species in that genus. We now learn that our commonest wheats belong to the genera *Agropyron* and *Aegilops* quite as much as they do to the genus *Triticum*. Were it possible to sacrifice convenience to accuracy, our bread wheats could more fittingly be designated as *Aegilotriticopyron*." (Since then, botanists have determined that the *Agropyron* weeds—quack grass—did not, in fact, contribute to modern wheat.)

The history written by the plants' own chromosomes was proved right in the end by plant breeder Sam McFadden. He crossed emmer wheat with *Aegilops squarrosa*, producing a hybrid with the chromosomes ABD. It was sterile, having an odd number of chromosomes. "Otherwise it looked like a primitive bread wheat," Anderson writes. Ernest Sears then took McFadden's sterile plants and treated them with colchicine, the chemical that doubles a plant's chromosomes. The result was a hexaploid wheat with the chromosomes AABBDD. It was "virtually identical" with spelt, says Anderson, and bred normally with ordinary European spelt varieties.

Millennia before genetic engineering was invented, how did humans succeed in transferring genes for cold tolerance and drought resistance from weeds into wheat? How did they break, not only the species barrier, but the presumably greater barriers between one genus and another? They did so by changing the conditions under which the plants grew. Their new technology, arable farming, allowed mutants with ordinarily lethal genetic changes not only to survive but to expand their range.

The difference between arable farming and other ways of encouraging plants to grow is plowing: the farmer breaks the soil, uproots all other vegetation, and sows the seed in bare dirt. There, protected from pests and predators, watered and weeded, a hybrid is at no disadvantage—provided it looks like the crop the farmer intended. As Anderson concluded after studying the patterns of hybridization of two common American weeds, "There were no hybrids in the wild, not because the two species were not cross-fertile, nor because bees did not carry pollen back and forth, but because in the strict interlocked economy of nature there was no room for something different. There was no niche into which they could fit." When he grew the two species side by side in a garden, hybrid weeds quickly appeared—and flourished. Likewise in the farmer's field the new wheats found their niche.

Mutant plants were also favored by other practices of early farming. In his Crop Evolution Laboratory at the University of Illinois, Jack Harlan found that three seasons of sowing and reaping were enough to eliminate seed dormancy. Producing wheats with seedheads that didn't shatter was an accident of technology. Harlan writes, "I have personally harvested wild-grass seeds by hand-stripping and with flint-bladed sickles, beater and basket, beater and boat . . . steel sickles, scythe, mechanical stripper, mechanical blower, binder, and power mower/swather followed by pickup combine and grain combine." He estimates that none of these methods, ancient or new, collects more than half the crop. Of the ancient ways, though, some seemed to be much less work, involving less stooping. For instance, he notes, "Having used both sickle and beater, I have long wondered why the sickle was ever preferred." Although archaeologists have not yet answered his question, they have determined the result of that switch.

Gordon Hillman and Stuart Davis of University College London experimented with planting wild wheat on the scale they thought the first farmers might have used, then calculated how long it would take for domestic wheat to appear. By domestic they meant a mutant with a seed spikelet that did not shatter in a breeze. In a 2- to 4-acre plot it would take 5 to 6 years. In a tiny plot only 30 yards square it would take 10 to 20 years. But, they warn, "domestication would have occurred only if the crops were harvested while partially ripe (or near ripe) by means of sickle reaping or uprooting." Although beating the grain into a basket is quick and requires less stooping, it leaves on the stalk those seed heads that don't shatter easily. These, write Hillman and Davis, "are stripped by birds, and even if their spikelets were to fall to the ground, their almost complete failure to penetrate ground litter and self-implant ensures their predation by rodents, birds, and ants." They contribute little to the crop the next year, even if the same field is replanted.

On the other hand, a farmer who uses a sickle to cut the stalks, then bundles the sheaves together to thresh out the grain later, loses the seeds from heads that do shatter easily. The seeds that make it home, to be stored and sown the next year, are those whose spikelets are held tight to the stalk by a sturdy rachis. "Crops sown from the harvested grain will reflect this increased proportion of tough-rachised forms, and the increase will continue, year on year, for as long as crops are always sown on new land from harvests taken from the previous year's new plots(s)," Davis and Hillman write. "Eventually the crop will be composed entirely of tough-rachised forms, and at this point domestication (in respect of the fixation of semitough rachis) is complete."

At about the same time that farmers in the Near East were reshaping grass and weeds into bread wheat, farmers in the New World were creating corn. Corn, or maize, known to science as *Zea mays*, may be the greatest feat of genetic engineering yet. While domesticating einkorn wheat took a "few simple changes," in the words of Jared Diamond, creating corn seems to have required a "drastic biological reorganization." Like wheat, corn depends on humans to sow its seeds, stuck tight as they are on its enormous ears, which themselves remain firmly attached to the stalk. But whence came this huge ear, "the span of a human hand and thick as an arm," as Columbus limned it to the king and queen of Spain in 1493?

That answer was a point of contention among plant scientists for most of the last century. Corn, unlike wheat, bears little resemblance to its wild relatives. Was its big-eared ancestor extinct? Or was corn, as some scientists asserted, an extraordinary, even catastrophic mutation of a wild grass called teosinte?

Botanists in the nineteenth century had classified teosinte as *Euchleana mexicana*: it looks so different from corn that no one suggested it should be placed in the same genus (*Zea*), not to mention the same species (*mays*). Teosinte looks like a grass. It has many branches and grows in clumps. Its ear, like an ear of early wheat, is a row of easily dispersed seeds, each enclosed in a hard hull or fruitcase. Crosses of corn and teosinte do exist in the wild, but they were not recognized as such until the late 1800s. They looked so different from either parent that they were classified as a third species, *Zea canina*.

Then in 1896 a Mexican agronomist named José Segura crossbred corn and teosinte, producing fertile plants that looked just like wild *Zea canina*. His work was noticed by G. N. Collins of the USDA's

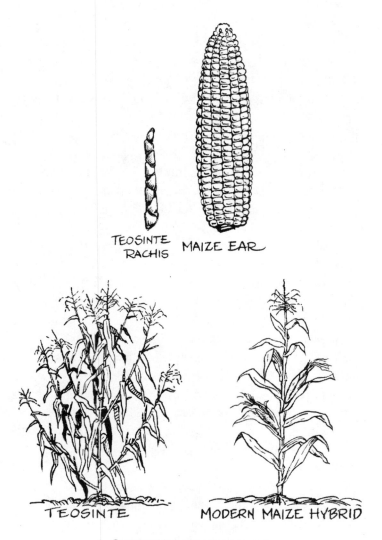

TEOSINTE RACHIS   MAIZE EAR

TEOSINTE   MODERN MAIZE HYBRID

Corn and its ancestor, teosinte

Bureau of Plant Industry, who traveled to Mexico in search of teosinte in 1919. Following the advice of Edward Palmer, an ethnobotanist at Harvard's Peabody Museum, Collins collected teosinte on the banks of an irrigation ditch in Durango, Mexico. He brought back both plants and seeds, which he grew until he had enough teosinte not only for his own research, but to share with other scientists, including Rollins Emerson of Cornell University. Emerson, a geneticist, was less interested in the history of corn than in the link between chromosomes and inheritance. Among his graduate students was George Beadle, who would share the Nobel Prize in 1958 for the "one-gene one-enzyme" hypothesis. Emerson asked Beadle to make hybrids between teosinte and maize. It proved easy to do: the plants are 100 percent interfertile. After studying the hybrids' chromosomes, Beadle and Emerson concluded in 1932 that maize and teosinte belonged not only to the same genus, but to the same species: the two plants shared even the order of the genes on each chromosome. Beadle and Emerson considered the puzzle of the history of corn solved.

Their solution did not persuade everyone. Other maize geneticists, particularly Paul Mangelsdorf, believed that the ancestor of corn was either extinct or had not yet been found. The differences in how maize and teosinte grew and reproduced were just too great. Mangelsdorf could not believe that teosinte, with its inedible seeds, could turn into the most productive food plant on the face of the earth.

In 1938 Mangelsdorf and Robert Reeves proposed a new theory of the origin of corn. Although cumbersome, it came to be widely believed. From Edmund M. East, Mangelsdorf's thesis advisor at Harvard University, they borrowed the idea that maize evolved from an extinct South American species. From Edgar Anderson, another student of East, they adopted the idea that teosinte came from a cross between maize itself and a grass in the genus *Tripsacum*. The diversity in contemporary corn, they said, could be traced back to this "infection" by *Tripsacum*. After many attempts they were able to germinate and grow a few, mostly sterile, hybrids between maize and *Tripsacum*. They also crossed maize-teosinte hybrids back to maize and analyzed these backcrosses, identifying a number of differences they attributed to the *Tripsacum* infection.

When two genomes that are very distant come together, the differences between them do not disappear. Telltale signatures persist in the DNA. Mangelsdorf's hypothesis would one day be testable. But in 1938 it was not, and Mangelsdorf and Beadle sparred for years about who was correct. (Some of their efforts were quite theatrical. According to Jack Harlan, to prove that teosinte was the same species as corn,

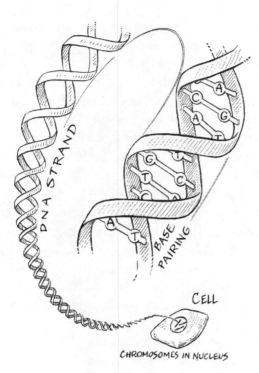

CHROMOSOMES IN NUCLEUS

DNA, the double helix

"George Beadle ground up fruitcases, seed and all, made tortillas out of the flour, and ate them.")

Beadle worked on a very different organism, the orange bread mold *Neurospora*, for most of his career and then served as president of the University of Chicago. Mangelsdorf became a professor at Harvard and influenced generations of archeologists through his collaborations with Harvard archaeologist Richard MacNeish, who studied early agriculture in Mexico. When Beadle retired he returned to the teosinte question. In 1971 he organized a teosinte hunt in Mexico to look for more wild maize relatives. Some of the teosinte seeds he collected on that expedition, from the Balsas River valley, found their way to John Doebley, now at the University of Wisconsin-Madison.

Doebley belongs to a new generation of geneticists able to ask much more detailed questions about the evolution of plants. With molecular markers and statistical methods, geneticists can now figure out which pieces of which chromosomes are responsible for the seem-

ingly huge differences between teosinte and maize. They can even estimate how many genes might be involved. Their methods rely on the
fact that DNA changes at a remarkably rapid pace.

It had long been known that DNA mutates. Each time a cell divides, its DNA is copied. During this copying process, mistakes—mutations—can occur. To make a copy, the long twisting DNA double
helix is first unwound: from resembling a spiral staircase, it begins to
look much more like a ladder. Then the rungs of the ladder, each made
of two linked bases, are separated. A new partner is found for each base
until there are two copies of the original DNA.

These copies are usually identical because the bases, known by the
first letters of their chemical names as A, T, G, and C, are very particular in their pairing. A pairs only with T; G pairs only with C. But the
copying mechanism does make mistakes, putting a C where an A
should go, for example. Such so-called "point" mutations are rare, occurring perhaps once in every million times a gene is duplicated. But
they do happen.

Point mutations do not always have an effect on the plant. Not all
of a plant's DNA is devoted to genes—to the code, the instructions, for
making proteins. And even a mutation in a gene doesn't always change
the protein that the gene codes for. Nor are these substitutions the only
kind of change that can happen to a DNA sequence. Frequently, the
molecular machine that copies DNA gets stuck and, in effect, stutters.
A short sequence such as AT becomes ATAT, then ATATAT, and so on.
These changes, which are quite common and usually have no effect,
are called simple sequence repeats or microsatellites. The reverse also
happens, when the DNA copier skips a set of repeats, and ATATAT
becomes ATAT or AT. Other kinds of mutations that add or delete
larger strings of bases are relatively frequent as well. Often they too are
harmless. A final source of change comes from transposons, familiarly
known as jumping genes. Maize and its ancestors are particularly rich
in these mobile stretches of DNA.

Because DNA is always changing in these small ways, the chromosomes of any two plants belonging to the same species differ in many
places. Plants like maize and teosinte, which have reproduced sepa-
rately for thousands of years, have accumulated many, many differ-

ences. Doebley and his students sought to link these differences in DNA to differences in the maize and teosinte plants, particularly in the structure of the all-important seed-bearing parts. They crossed maize to teosinte, then backcrossed the hybrid to its maize parent for many generations. When they compared the various offsprings' DNA (automated DNA sequencing machines and massive computer programs have made this task quite simple), they could identify even small lingering bits of teosinte DNA.

Emerson, Beadle's advisor, had suggested back in the 1920s that only a few genetic changes were needed to transform teosinte into maize. In 1992 Doebley and his colleagues came to much the same conclusion: changes in no more than five major regions of the genome made the difference. In two cases they could trace the change to an individual gene.

One was a gene that affects the glume, one of the structures that encloses the kernel. In teosinte the glume is hard, almost stonelike. It forms the fruitcase around the kernel. This stony fruitcase ensures that even if the kernel is eaten, it passes through the eater's digestive tract unscathed. But the plant's success in thus scattering its seeds is a nutritional failure for humans. The hardness, size, and curvature of the glume, Doebley and his colleagues found, is controlled by a gene named *teosinte glume architecture,* or *tga1.* (Genes are often named for the way a mutation in the gene affects the plant. Two mutations giving the same kind of phenotype, or form, even if they involve different genes, are given the same name, but different numbers are used in their abbreviations.) The hardness of the teosinte glume is due partly to silica deposits in its outer cells (silica also makes the surface of teosinte kernels shiny) and partly to lignification (some glume cells become filled with lignin, the same substance that makes wood hard).

Maize kernels do not have a stony fruitcase. In a plant that has the maize version of *tga1,* silica is deposited in only a few glume cells. The cells are also less lignified. Finally, in maize the glumes grow more slowly, so they do not fully enclose the kernel. A change in this one gene moved teosinte a long way toward becoming a useful food plant. It was quite bad for teosinte, however, because its kernels were no longer as resistant to destruction. The only way the mutation would persist is if people made sure some of the seeds were planted. George Beadle believed that people must already have been harvesting, grinding, and probably cooking teosinte seeds when the mutation occurred. They were most likely farmers as well, because squash was domesticated even longer ago than maize. To these early farmers, the change to softer seeds probably just reduced the amount of work it took to make the seeds edible. Hugh Iltis, on the other hand, has suggested that teosinte plants were cultivated at first because their stalks are sweet, much like sugar cane, and that it was the *tga1* mutation that first made the seeds useful as human food.

The second gene Doebley's team studied was named *teosinte branched* or *tb1*. Teosinte plants have lots of long side branches that make them look like bushy grasses. Each side branch has a tassel, a male pollen-producing flower, at its tip. Teosinte's female flowers, which become the ears, are produced by secondary branches that grow from these main branches. Maize, on the other hand, is generally not branched. It has one main stalk with a pollen-producing tassel at the very top. Its side branches are very short and end in ears, the female flowers. Much of the difference in the way the two plants look and grow is due to the *tb1* gene. The maize form of the gene keeps the lateral shoots from growing long, converting the bushy shape into the slim, single-stalked maize. It telescopes the side-branches into ears and surrounds them with layers of leaves—the husks—growing close together. Finally, it converts what would have been male flowers, the pollen-producing tassels, into female ears.

Hugh Iltis had predicted this mutation in 1983, when he was director of the herbarium at the University of Wisconsin-Madison. He called it "the catastrophic sexual transmutation." It was, he wrote, "a gross and sudden quantum evolutionary emergence of a 'hopeful monster," turning a male part into a female part. Iltis has been described as "the Sir Richard Burton of the plant world, scaling the Andes in search of a wild potato." His theory of how teosinte turned into corn was "as flamboyant as his person," but by 2002 he had reconsidered the "monster" part of his theory. At a botany conference he noted that his theories on how to transmute teosinte tassel spikes into maize ears were "unnecessary explanations of common phenotypic variability. This was well understood but its implications totally missed by this author. To quote Mark Twain, 'The eye cannot comprehend when the imagination is out of focus.'"

Interestingly, both the *tga1* and *tb1* genes code for proteins that control the expression of a group of other genes. For *tb1* it is not the structure of the protein that makes the difference, but how much of it is made. The part of the *tb1* gene that encodes its protein did not change much as teosinte evolved into maize. But maize plants produce twice as much of the *tb1* protein as teosinte plants.

If several such mutations were needed to create corn, then the plant at some point in its evolution might have met a bottleneck. An evolutionary bottleneck is a time when the size of a population is vanishingly small. It consists of only a few plants, each carrying the crucial mutations. All of the maize plants throughout the world would be descended from these few plants. This prediction, too, can now be tested. A large population of individuals has in the aggregate a great many differences in its DNA. Based on how frequently these differences arise, scientists can create a molecular clock. Using a molecular clock for maize, Brandon Gaut and his colleagues, then at Rutgers University, estimated how many generations and how many founding parents it would take to account for the variability in modern corn. Their results were consistent with a bottleneck of just 10 generations and a founding population of 20 individuals.

The traits that distinguish corn from teosinte seem to come as a package. In teosinte the rachis that holds the seeds always has two rows, it always has single spikelets holding its kernels, and it always shatters easily. The maize ear, by contrast, always has many rows and pairs of spikelets, and it does not shatter. Yet several genes contribute to this cluster of traits, to rows, spikelets, and shattering. One way these traits could all change at once is if there were a single controlling gene: A mutation in the controlling gene might then change how all the genes under its control were expressed, affecting the whole cluster of traits at the same time. Or there could be another explanation: differences in the genes could accumulate undetected—as cryptic changes, hidden variation—until, at some critical point, one final change reveals all the others.

To see if there was cryptic variation in teosinte, Doebley and his student Nick Lauter made a hybrid between two different types of teosinte (subspecies *mexicana* by subspecies *parviglumis*), then crossed that hybrid to maize. They looked at seven traits in the hybrid—such as an easily shattering ear—that couldn't be detected in either one of the parent teosintes. Did either of the teosinte subspecies, they asked, have an effect? For four of the seven traits, they found, genes from both teosinte parents made the maize-teosinte hybrid more maize-like. For the other three, one of the teosinte parents made a maize-like contribution, the other one did not. There was, they concluded, cryptic variation in teosinte, hidden mutations that were suddenly revealed, switching the plant onto a new evolutionary track.

Finally, they asked, had this sudden change happened just once or did it happen many times in many places? Because maize is such an extraordinarily variable plant, both in its appearance and in its genome, scientists had conjectured that teosinte had been transformed into maize many times. This question can be answered by constructing family trees based on the genetic similarities and differences among modern corn varieties and various types of teosinte. If maize arose from different types of teosinte multiple times, we should be able to derive several family trees, tracing the domestication events back to the particular variety of teosinte that gave rise to each one. If it happened only once, all modern corn varieties should fit into a single family tree tracing back to a single teosinte ancestor. Doubley's group published the results of this experiment in 2002. They had screened a hundred different genes using microsatellites, the small repetitive sequences that change so rapidly. Their results were unequivocal: all modern corn belongs in a single family. It was domesticated only once.

Knowing how fast genetic differences arise, and how many there are today, they could also estimate when this event happened. Moreover, knowing where the descendents of the genetically closest teosinte live today, they could pinpoint where it happened. Maize, they concluded, arose from teosinte of the subspecies *parviglumis* in the Balsas River basin of southern Mexico between 5,000 and 13,000 years ago.

Until recently the oldest archaeological evidence of domesticated corn came from cobs found in the San Marcos Cave in the Tehuacan

Valley in southern Mexico. These cobs are roughly 4,700 years old. In 2001, however, cobs from the Guila Naquitz Cave, excavated in 1966, were redated by accelerator mass spectrometry and found to be more than 6,200 years old. While these early cobs do not look much like our huge modern corncobs, they look even less like ears of teosinte. All three fossilized cobs have kernels that are tightly attached, unlike teosinte. While two of the fragments have only two rows of kernels, like teosinte, one has four, more like modern corn. Moreover, in these ancient cobs the grain-bearing spikelet is set perpendicular to the stem, as it is in maize but not in teosinte. Finally the glumes of the kernels appear to be flexible, like those of maize, suggesting that the *tga1* gene had already mutated and the kernel had lost its stony fruitcase.

The Guila Naquitz Cave is in southern Mexico, 400 to 500 kilometers east of the Balsas River area where corn's closest teosinte relative lives today. The age of these earliest cobs pushes the origins of corn back to well within the estimate arrived at by the molecular biologists. The new maize varieties traveled from hand to hand with incredible speed—on an evolutionary time scale—creating what might be regarded as an early "Green Revolution." Recent molecular analyses of the *tb1* gene show that by 2,000 years ago a crop containing the maize version of the gene was already being grown as far north as New Mexico.

For corn, domestication was wildly successful. Humans remade the landscape, clearing forests and plowing-under the prairies to plant more corn. The first Europeans to reach the New World, writes Margaret Visser in her book *Much Depends on Dinner*, were "staggered by the size of the maize plantations: Diego Columbus, Christopher's brother, said he walked 29 kilometers (18 miles) through an immaculate field of corn, bean, and squash mounds, and never came to the end of it." Columbus brought maize back to Spain; now it grows on every continent, covering 80 million acres of America alone. While its teosinte relatives remained in a very small geographical area, corn can now be found at latitudes from 50° N to 40° S, and at elevations from sea level to 12,000 feet.

The same is true for wheat. "Wheat is not quite as adaptable as man, who exceeds all other species in the range of environments in which he can survive," writes Felipe Fernandez-Armesto in his history of food, *Near a Thousand Tables*, "but it has diversified more dramatically, invaded more new habitats, multiplied faster, and evolved more rapidly without extinction than any other known organism." It grows on more than 600 million acres around the world. Fernandez-Armesto predicts that far in the future, when historians look back on our age, "They will classify us, perhaps, as puny parasites . . . whom wheat cleverly exploited to spread itself around the world." Yet world dominance for wheat and corn came at a cost: domestication meant that each plant would henceforth and forever be a ward of humans, dependent on them for its survival—just as human societies soon became dependent on their crops.

The new foods could be easily harvested and stored. Barns (originally, places to store barley) and granaries (for any grains) were built—which is one possible explanation for the circular tower found in ancient Jericho. To protect their food stores, once-nomadic tribes settled permanently. Towns developed in which full-time specialists—potters, weavers, dyers, metallurgists, masons, priests, and others—could practice their arts and professions supported by the surrounding farmers and herdsmen. "Where agriculture was invented," writes Joel Cohen of the Laboratory of Populations at Rockefeller University, "local populations grew ever so slightly faster. Whether the invention of agriculture enabled the population to grow faster or a faster-growing population was driven to devise agriculture, or both—these remain questions for speculation."

There was, in any case, a link between the two. "In all parts of the world where adequate evidence is available," writes Jared Diamond in *Guns, Germs, and Steel*, "archaeologists find evidence of rising densities associated with the appearance of food production." Agriculture is, he explains, "an autocatalytic process—one that catalyzes itself in a positive feedback cycle, going faster and faster once it has started. A gradual rise in population densities impelled people to obtain more food, by rewarding those who unconsciously took steps toward producing it."

When farming began, 10,000 to 50,000 years ago, the world human population was between 4 and 10 million. Then it began to grow. By the time of Christ it had risen to between 100 and 300 million. By 1700 it had reached 600 million. Cities, kingdoms, and empires were built on the growing of grain.

People in early agricultural societies were not necessarily healthier than their hunter-gatherer relatives. Farming did not provide nearly as varied a diet as hunting and gathering had. Crops failed, stored grain grew moldy. Studies of early skeletons show that the first farmers were less robust than their hunting ancestors. They suffered from nutritional disorders, infections, and diseases that their ancestors had escaped. The amount of protein and the number of calories each person ate declined for everyone except members of the privileged class. Yet the farmers' children might have been more numerous. Grains, pounded into gruel, notes anthropologist Mark Nathan Cohen, "would have simplified the problem of feeding the very young whether or not it improved nutrition." Corn-fed children can be weaned earlier and their mothers (if well fed) can bear the next child sooner. And if the birth rate can exceed the death rate, even a poorly nourished people can multiply.

As Diamond remarks, "Some productive hunter-gatherer societies reached the organizational level of chiefdoms, but none reached the level of states: all states nourish their citizens by food production." He adds, "Food production was indirectly a prerequisite for guns, germs, and steel. Hence geographic variation in whether, or when, the peoples of different continents became farmers and herders explains to a large extent their subsequent fates."

Part of that geographical variation has to do with the third plant trait people changed through their early efforts in genetic modification. Although they did not understand what they were doing in terms of genes and genomes, by choosing which seeds to eat and which to sow the first farmers were identifying the slightly different versions of genes, the alleles, that made a plant more useful to humans.

The crucial alleles in a crop control three characteristics: the size of the seed, how and when it is dispersed, and the plant's dependence on day length. How long it takes for a plant to flower and set seed determines where it can grow. Yet exactly when a plant flowers in the summer—and it must flower to set seeds—depends on how long the days are. Plants have many proteins that absorb light. Among them are phytochromes, which absorb red light, and cryptochromes, which absorb blue light. These proteins interact with each other to sense both the quality of light—how much blue and how much red—and to monitor the length of the day. Only when both are right does the plant flower.

On the equator, the length of the day is constant all year round. Because the earth's axis is tipped, the farther north or south you go, the greater the difference between the length of a summer day and that of a winter one. Plants read these cues precisely. To make a plant flower outside its original zone means tampering with the genes required for sensing the length of a day. This kind of genetic modification is easy, though it might take centuries: when people grow crops in a new environment, they can harvest and replant the seeds from only those plants that flower. The formation of potato tubers is also controlled by day length. This effect might well be why Parmentier was able to grow potatoes in Paris just before the French Revolution when his predecessors had failed.

There are more than seven species of cultivated potato. Some species, like wheat, are polyploids, with three, four, or five sets of chromosomes instead of a single pair. Like corn, they and their wild relatives hybridize very easily. Potatoes can, as the eighteenth-century French feared, be poisonous. Both wild and domesticated potatoes contain bitter chemicals called glycoalkaloids which can reach toxic levels. "In the cultivated potato," Jack Harlan writes, "strains have been selected that are relatively safe, but even today, in areas far removed from the Andes, certain clones under some conditions can be dangerous."

In the potato's homeland the problem is more acute. "In the Andes, strains selected for low toxicity often cross with wild and weedy races and toxic tubers are produced: the local people must somehow live with the poisons." Many of them use an elaborate freeze-drying technique to get rid of the toxins. Others dip potatoes in a certain kind of clay which, like Kaopectate, protects the stomach. Chemical ecologist Timothy Johns concludes, "Undoubtedly potatoes were an unpredictable and dangerous resource until some way was found to eliminate their potential toxicity."

Modern potatoes, from the Idaho to the All-Red, are all varieties of a single subspecies, called *tuberosum*, of the species *Solanum tuberosum*. This subspecies originated in Chile. The first potato found by the Spanish explorers and brought to Europe, however, was a different subspecies of *Solanum tuberosum*. It was called *andigena* because it came from the Andes. *Andigena* potatoes were described by Juan de Castellanos in 1537 as "white and purple and yellow, floury roots of good flavor, a delicacy to the Indians and a dainty dish even for Spaniards." They did not grow well in Europe. Because they came from a high altitude but a low latitude, writes plant physiologist Lloyd Evans, "*andigena* potatoes were well adapted to the cool conditions but not to the long days of European summers. Although it was not understood at the time, the long days of summer prevented tuber formation until close to the autumn equinox, leaving little time for tuber growth." It took 250 years of selection by European gardeners and botanists—who grew potatoes as ornamental bushes, beautiful but "savage-looking" with their hairy stems and blue-purple flowers—before Parmentier had a plant he and his fellow Parisians could eat. Unfortunately, Parmentier was unable to persuade his countrymen to do so until hunger had led to revolution.

"The tuber ran smack into centuries-old customs that decided two of the most basic questions of peasant life, what to put in the soil and on the plate," writes Larry Zuckerman in *The Potato: How the Humble Spud Rescued the Western World*. "Existence was fragile enough without letting in what was new and uncertain, even though what was new might help. The fear of change ran so deep that people were ready to die rather than alter their ways."

3 THE POWER IN THE EARTH

It is deplorable but true that many agricultural scientists in some advanced countries have renounced their allegiance to agriculture for reasons of expediency and presumed prestige. . . . Let the individuals live with their own motivations; let them serve science and themselves if they wish. But the institutions have the moral obligation to serve agriculture and society also.

—Norman Borlaug (1970)

In 1798 the English economist and cleric Thomas Malthus laid out what he called the immutable laws of human existence: "First, That food is necessary to the existence of man. Secondly, That the passion between the sexes is necessary and will remain nearly in its present state."

Passion, in Malthus's time, led inexorably to children (he had no inkling of birth control), and thus to a population that, when unchecked by famine, disease, or war, increased in what Malthus called a "geometrical ratio." Assuming that each married couple would have 4 children and that all of them would marry, 2 would become 4 would become 8 would become 16 would become 32 would become 64. The "power in the earth" to produce the necessary food for those people, however, Malthus reasoned, can at best increase in what he called an "arithmetical ratio" over the same timespan: 1, 2, 3, 4, 5. . . . Following this reasoning, Malthus argued that in a couple of centuries there would be 512 times as many people, but only 10 times as much food.

Even if "by breaking up more land and by great encouragements to agriculture" every acre in Britain were to be made into a garden, Malthus concluded, the country would not, within 50 years, be able to feed its people.

Malthus was extremely influential. Charles Darwin's principle of natural selection, for instance, was a result of his reading Malthus's *Essay on the Principle of Population.* At the same time, Malthus acquired a reputation for being hardhearted. He blamed the declining quality of life in Britain on the irresponsibility of the lower classes—they simply had too many children. He believed that government aid encouraged them to have more children than they could support. He thought that only those who worked should eat. A Malthusian worldview remains synonymous with the gloomy belief that only famine, disease, and war can control population and solve the world's food problems.

But birth control was not the only concept Malthus's essay did not take into account. On the cusp of the nineteenth century, several scientists were struck by the idea of increasing what Malthus called the "power in the earth" to produce food. They began to look at agriculture scientifically. The German botanist Julius von Sachs was one of this group. An expert on the tiny hairs on a plant's roots, Sachs invented several devices to quantify how those hairs took up water and nutrients. The secret to his "artificial plants," grown in water, freed of soil, was knowing what to put in the water: that is, understanding the chemistry of plant life.

Ever since 1772, when English chemist Joseph Priestley grew mint inside a sealed jar and observed that even months later the air within would "neither extinguish a candle, nor was it at all inconvenient to a mouse," scientists had known that plants give off oxygen. That the same process, photosynthesis, also provides the carbon of which plants are mostly made was suggested in 1779 and proved "with classical completeness," says Sachs, in 1804. In that year Theodore de Saussure showed that in the presence of sunlight the green parts of plants take in carbon dioxide and water and turn them into sugar and other carbohydrates, giving off oxygen.

Yet, Sachs notes, "For 40 years subsequently, almost inconceivable misapprehensions again obscured the clearly established fact" that plants literally make food out of thin air. The carbon and other elements in plants, people argued, must be different from the elements in rocks and air. In living things, they thought, the elements must contain a vital force. In 1828 German chemist Friedrich Wöhler believed he had disproved the vitalist notion when he synthesized urea, considered vital because it was excreted by animals. "I must tell you," he wrote to a friend, "that I can make urea without the use of kidneys either man or dog. Ammonium cyanate is urea." And nitrogen, therefore, was nitrogen.

Still, it was not until the 1840s, when Wöhler's friend and colleague Justus von Liebig published his *Organic Chemistry and its Applications in Agriculture and Physiology*, that these advances in plant science began to affect farming. Liebig's book, which focused on soil fertility, sold thousands of copies in America alone, while his letters to agricultural magazines reached many more farmers worldwide. When the Civil War began in 1861 Liebig was far better known around the world than Abraham Lincoln.

The first application of agricultural science was fertilizer. Before 1842 English farmers spread on their fields saltpeter from India, nitrate of soda from Chile, guano from Peru, slag (a waste product from the iron foundries), gypsum, lime rocks, and bones. Bones and lime rocks contain the phosphate plants need, but they are still poor fertilizers: their calcium phosphate dissolves poorly in water and so is hard for a plant to absorb. James Murray, an Irish doctor with a calling toward chemistry, found that acid converts calcium phosphate into calcium hydrogen phosphate and calcium dihydrogen phosphate, both of which dissolve readily. Murray mixed a paste of acid-treated bones or rocks with sawdust and bark to produce a slow-release fertilizer. He published his *Advice to Farmers* in 1841 and obtained both Scottish and English patents in 1842.

Sir John Bennett Lawes of England, who had also experimented with soaking bones and rocks in acid, bought Murray's patent in 1846 and in time came to be known as the father of the fertilizer industry. His superphosphate of lime was thought to be such an improvement that, "as a testimonial," wrote Evan Pugh, a young American chemist, the country's farmers "gave him his choice, a *Laboratory* or its value of

Plate," silver plates and tableware being a common reward at the time. He chose the laboratory, which was built on his estate at Rothamsted, England, and became a model for agricultural experiment stations in both England and the United States.

Pugh came to Rothamsted in 1857, fresh from studies in Germany with Wöhler and Liebig. There he occupied himself with the "nitrogen question." Would nitrogen added to the soil as fertilizer increase the yield of a crop, or did plants take in and fix enough nitrogen from the air? An experiment in Paris, by chemist Jean-Baptiste Boussingault, had shown that plant nitrogen came from soil nitrates and ammonia, suggesting that nitrogen fertilizer was a good idea. Boussingault's data, however, were found to be faulty, and rival work by another French scientist, Georges Ville, claiming that plants took nitrogen directly from the air, was taken up and popularized by the great Liebig.

Pugh sided with Boussingault. At Rothamsted, he wrote, "They have supplied me with about $500 worth of apparatus and we have been doing up the subject on a scale unprecedented." In sterilized soil sealed under glass Pugh tried to grow three kinds of cereals and three kinds of legumes. Into the glass containers he pumped air, first washing it through both acid and potash, then adding back carbon dioxide. None of the crops grew. His results, he concluded, "indicate a confirmation of Boussingault. The evidence accumulates that Ville is an ass." When he asked a visiting French scientist "how he accounted for Ville's plants growing as they did," the man answered, "*Il a ajouté sans doute*" ("Without a doubt he has added something"). Neither he nor Pugh was aware that what Ville might have added, simply by neglecting to sterilize the soil, were bacteria that did fix nitrogen out of air. In Germany, through his water-culture experiments, Sachs easily replicated Boussingault's and Pugh's results, and even Liebig was convinced. The nitrogen fertilizer industry—now of such importance to agriculture, both because of the greater yields it allows and the environmental problems it has caused—was born.

Chemistry was not the only nineteenth-century science to change— and be changed by—agriculture. In 1859 English naturalist Charles

Darwin published *The Origin of Species*, beginning a scientific debate that would last well into the 1930s and, in some circles, until today. In developing his theory of natural selection Darwin drew not only upon his observations of finches and tortoises in the Galapagos islands, but also on his knowledge of domesticated plants and animals. "We can not suppose that all the breeds were suddenly produced as perfect and as useful as we now see them," he wrote. "Indeed, in many cases, we know that this has not been their history. The key is man's power of accumulative selection: Nature gives successive variations; man adds them up in certain directions useful to him. In this sense he may be said to have made for himself useful breeds."

As an authority, Darwin cites William Youatt who, he says, "was probably better acquainted with the works of agriculturalists than almost any other and who was himself a very good judge of animals." Youatt, Darwin says, "speaks of the principle of selection as 'that which enables the agriculturist, not only to modify the character of his flock, but to change it altogether. It is the magician's wand, by means of which he may summon into life whatever form and mould he pleases.'"

It was not *The Origin of Species*, but a later book of Darwin's that inspired Luther Burbank to take up the magician's wand. The man who would become known as the wizard of horticulture was 19 when he read Darwin's *The Variation of Animals and Plants Under Domestication*, published in 1868. "It opened a new world to me," he said simply. He was a truck gardener at the time, working a small plot of land next to his mother's house and trucking his vegetables to New York City to sell. Upon reading Darwin he decided to distinguish himself from the rest of the vegetable carts in the market by creating new plants.

By 1873 he had built, as he put it, the Burbank potato, also called the Idaho potato, that russet-skinned, oblong standard used to make McDonald's french fries. Potatoes are most often clones: they usually reproduce through tuber sprouts, so each potato plant is genetically identical to its parent. Yet sometimes a potato plant sets fertile seed. Discovering one of these rare seedpods at the edge of his garden, Burbank grew the 23 seedlings and selected 2. From their offspring he selected again, then sold his new Burbank variety to a Massachusetts seed salesman. With the $150 he earned he moved to California. There,

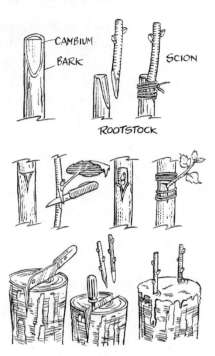

The art of grafting

beside "the world's largest wheat field, an 80-mile-wide strip along the banks of the San Joaquin River," says his biographer, Peter Dreyer, he envisioned orchards. He imported berries and plums and nuts from Japan, Panama, France, Chile, Argentina, Mexico, and Spain. He grew hundreds of stock trees, each holding dozens of grafts.

The technique of grafting had been known since ancient times. Yet in early America, grafting—just like molecular techniques today—was condemned as unnatural, as interfering with God's plan. The popular sect called the Swedenborgians preached that all material things (rocks, plants, animals, and people) were reflections of the spiritual world and should not be tampered with. John Chapman, better known as Johnny Appleseed, believed that grafting violated the divine essence of an apple tree. He said: "They can improve the apple in that way, but that is only a device of man, and it is wicked to cut up trees that way. The correct method is to select good seeds and plant them in good ground and God only can improve the apple." When he gave away apple seeds, he insisted that the trees be allowed to grow freely so that they could express their spiritual reality. Orchardists, whose livelihood depends on a predictable crop, did not take Chapman's seeds. Orchardists, writes Sue Hubbell in *Shrinking the Cat: Genetic Engineering Before We Knew About Genes*, plant "proven grafted stock and never, never, never save apple seeds to plant unless they have a speculative, adventuresome, experimental set of mind." Because of the "perversely complicated" genetics of the apple, every tree that grows from a seed is a new variety—and most of them are commercial failures.

Fortunately for fruit lovers, the divine essence of the orange tree was not a consideration when the seedless navel orange was discovered as a sport or bud variation on a seedy variety in eastern Brazil in the early 1800s. In 1869 12 navel orange trees arrived at the U.S. Department of Agriculture's new greenhouse in Washington, D.C., the gift of a Presbyterian missionary. To create the 12 new trees, horticulturalists had slit open the bark of a number of normal, seedy orange trees and carefully inserted into each wound a bud taken from a branch of the seedless sport. If the bud took, the branches of the stock tree were broken off to send all growth into the one new seedless branch. Every navel orange is thus a clone, a genetically identical twin, of that first Brazilian bud. In 1873 two such clones were sent to Riverside, California; from them sprang the California citrus industry.

Likewise, by such expedients as grafting dozens of different plums onto each one of his hundreds of stock trees (some of them almond trees), Burbank built up the California plum industry. Eleven of the plum varieties currently popular among California growers are Burbank's.

Using tweezers and scalpels to emasculate plants, and paintbrushes, not bees, to transfer pollen to the female ovules, Burbank also made thousands, if not millions, of crosses. He crossed peaches with almonds, plums, and apricots. Plums he crossed with almonds and apricots, quinces with apples, and potatoes with tomatoes. To make his hybrid berries, he claimed to have crossed 37 different species in the genus *Rubus*. He was "the Henry Ford of the art," Dreyer writes. "He brought mass production to hybridizing, raising thousands of seedlings to obtain a single improved variety."

Burbank destroyed his rejects in huge bonfires, lighting 15 in one year, with one fire alone destroying 65,000 hybrid berry bushes. Consigned to the fires were several hundred failed attempts at a white blackberry. Having found a yellowish berry (grown under the hyperbolic name Crystal White), Burbank decided to cross it with a black variety to see if, paradoxically, that would whiten the fruit. The first generation was still black. But when those plants were bred, a few of the offspring did bear whiter fruit. Seeds of these whitish berries were planted and, out of several hundred plants, five bushes bore nearly colorless fruit. The best of these became the Iceberg blackberry. The rest were burned.

Explaining his techniques in his 1893 catalogue, *New Creations*, Burbank announced, "We are now standing just at the gateway of scientific horticulture." His fruits and nuts and berries "are more than new in the sense in which the word is generally used; they are new creations, lately produced by scientific combinations of nature's forces, guided by long, carefully conducted, and very expensive biological study. Let not those who read suppose that they were born without labor; they are not foundlings, but are exemplifications of the knowledge that the life-forces of plants may be combined and guided to produce results not imagined. . . . Limitations once thought to be real have proved to be only apparent." In a speech in San Francisco in 1901, he elaborated. Botanists, he said, had once "thought their classified species were more fixed and unchangeable than anything in heaven or earth that we can now imagine. We have learned that they are as plastic in our hands as clay in the hands of the potter or colors on the artist's canvas, and can readily be molded into more beautiful forms and colors than any painter or sculptor can ever hope to bring forth."

Burbank's was the only plant breeder's name to enter *Webster's Dictionary* as a verb, "to burbank: to modify and improve (plants or animals), especially by selective breeding." Yet he was not the only plant breeder feverishly introducing new varieties. Just as he imported his source material from Japan, Panama, France, Chile, Argentina, Mexico, and Spain, other breeders in America were using seeds and plants from Germany, the East Indies, Poland, Russia, and Ethiopia.

Since 1838 the U.S. Navy had sponsored official plant exploration expeditions, bringing home new types of vegetables, barley, rice, beans, cotton, persimmons, tangerines, roses, and wheat, which were then burbanked to produce varieties suitable for growing in America. The United States Commissioner of Patents had gained permission—and congressional funding—in 1839 to collect and distribute seeds, plants, and agricultural statistics. These seeds were, by the late 1880s, being quality-tested by the U.S. Department of Agriculture (USDA), which took over from the Patent Office in 1862, and distributed seeds and plants to breeders and farmers through the state agricultural experiment stations, which were modeled on England's Rothamsted.

The turn of the twentieth century would see "the golden age of plant hunting," with 48 USDA-sponsored expeditions in the next 25 years. These expeditions, and the foreign varieties they introduced, completely changed the crops American farmers planted and grew. Yields, however, remained static between 1860 and 1900. "The new varieties that flowed in an ever greater stream from public researchers were not raising average yields, but were permitting extension of production into new areas," noted rural sociologist Jack Kloppenberg in his book *First the Seed: The Political Economy of Plant Biotechnology, 1492-2000*. "Advances in plant breeding served to *maintain* levels of yield that might otherwise have declined." It was the scholarly insights of Gregor Mendel, not the brute-force approach of Luther Burbank, that increased crop yields without bringing more land under cultivation.

Mendel too had been inspired by Darwin: he bought every book Darwin wrote. What were these successive variations Darwin spoke of, he wondered? Mendel was an Augustinian monk; he lived in Brno, or Brünn, which was then part of the Austrian Empire and is now in the Czech Republic. In the monastery garden, Mendel grew peas with round, yellow seeds and ones with wrinkled, green seeds. He plucked the male stamens from a newly opened flower of the round and yellow type, and brushed onto the female stigma pollen he had collected from a flower of the wrinkled and green type. Then he charted the results: how many round, how many wrinkled, how many yellow, how many green.

Mendel was not interested in just plants. He collected bees from

Europe, Egypt, and America and tried to hybridize them. He kept colonies of mice and crossbred different strains. In 1866 he published his laws of genetic inheritance. But his work was not to have an effect on agricultural science until the twentieth century, when his "factors of heredity" were named "genes." "His mathematical approach, comparing the frequency of a trait in parents and offspring, was then unfamiliar in biology," notes one scientist, "and so his discovery was ignored."

The research that led to the rediscovery of Mendel was driven itself by curiosity about Darwin's notion of species. In 1886 Dutch botanist Hugo De Vries had noticed the great variety of evening primroses growing on the coastal dunes of his native Holland. To explain their diversity in terms of Darwin's theory, he formulated the idea of mutation (coining the word from the Latin *mutare*, to change), to mean an abrupt change that leads to a new species. He began breeding plants and worked out the pattern of inheritance of certain traits. Before publishing his results, he surveyed the earlier literature on heredity—and came across Mendel's "Experiments with Plant Hybrids" from 34 years earlier.

This time Mendel's work was not ignored. Although the concept of a gene was not yet clearly defined, the name "genetics" soon came to be applied to the Mendelian study of heredity. At last, said English biologist William Bateson, speaking at the Second International Conference on Plant Breeding and Hybridization held in New York in 1902, the plant breeder "will be able to do what he wants to do instead of merely what happens to turn up." Before, hybridization gave only "a hopeless entanglement of contradictory results" and "incomprehensible diversity." Now, using Mendel's laws, breeders would have the power of control. "An exact determination of the laws of heredity," Bateson said, "will probably work more change in man's outlook on the world, and in his power over nature, than any other advance in natural knowledge that can be foreseen." Bateson's statement was remarkably prescient: Mendel's laws did lead, if indirectly, to contemporary molecular techniques for plant improvement.

One of the few plant breeders not excited by Mendelism was Luther Burbank, who learned about it just after releasing another of his triumphs, the Shasta daisy. Mendel had little to offer the 50-year-old breeder, unequaled in the variety of fruits and flowers he had developed. Burbank's trial-and-error method was efficient enough for him. He stood "between the small-scale farmer-breeders who through most of history slowly but steadily built up mankind's stock of useful plants, and the modern scientific hybridizer," writes his biographer. In place of theory he had "a special gift of judgment," according to De Vries, who had made a point of going to see Luther Burbank's gardens.

Another geneticist who wondered about Burbank's secret was George Harrison Shull. Shull worked at the Station for Experimental Evolution, established with the immense sum of 10 million dollars from the Carnegie Institution of Washington at Cold Spring Harbor, New York, in 1904. Charles Davenport, the director, was an advocate of biometrics, the application of statistics to evolutionary theory and an alternative to Mendelism. Biometricians measured minute differences between closely related organisms in order to explain just how evolution worked: why organisms were like their parents, yet different enough to support change. Davenport studied chickens (and later, humans, charting the inheritance patterns of hair, eye, and skin colors; the work would embroil him in the eugenics movement which sought to limit childbearing to the "fittest"). Davenport gave Shull, a former student of his, two tasks: to explain the science behind Burbank's successful hybridizations, and to prepare a demonstration, using corn, of the simple inheritance patterns that Mendel had noticed in peas.

From 1906 to 1910 Shull spent part of each summer in California observing and questioning Burbank. They were not a good match. "You say, 'He is always impatient of a conversation in which he does not do all or nearly all the talking,'" Robert Woodward, president of the Carnegie Institution, wrote to Shull. "Now, singularly enough, this is precisely the remark he has made to me concerning you." Shull was a scientist; Burbank had "the work habits of an artist," his biographer notes. His recordkeeping system "was haphazard in the extreme." He was secretive and suspicious, searching his workers' pockets for seeds each evening as they left the job. Shull eventually gave up in frustration, never publishing his report. Burbank's success was due not to any scientific theory, he concluded, but to his "vivid imagination," "persistency," "concentration," "sensitivity to variation," and "exceptionally keen eye for what the market wanted." When asked, for instance, if he could breed a good-tasting decaffeinated coffee hybrid, Burbank replied that "it would involve years of experiments in the tropics." Besides, he added, "Would coffee be used, except for the exhilaration accompanying the caffeine? I think it would, but this is for someone else to decide." A few years later, a method was patented to extract caffeine from unroasted beans using steam and the solvent benzol. The resulting product was sold in France as Sanka (*sans caffeine*).

Burbank's hybrids had nothing to do with his own work, Shull felt, in the new field of genetics. Even when both men were experimenting with corn, Burbank crossing Indian corn with teosinte, which he believed was corn's ancestor, and Shull attempting to demonstrate Mendelian inheritance, the two didn't discuss their work. "How amazing," wrote geneticist and historian Bentley Glass, "that there was really no meeting of minds in the very area in which it might most reasonably have been expected!"

In New York, Shull crossed round-kernelled corn from his father-in-law's farm in Kansas with wrinkled corn from a farm near the lab. Corn, like Mendel's peas, has both male and female parts on one plant, though for corn the two sexes are far apart while for peas they are in the same flower. To cross two corn varieties, Shull plucked off the tassels of one variety, letting the other provide the pollen to fertilize the female flowers; he was thus assured that any ears resulted from a cross between the two.

The Kansas corn had full, round kernels because those kernels held more starch: being a long and bulky polymer, the starch filled the space inside the kernel completely. The New York kernels were wrinkled and collapsed because of a mutation that blocks the conversion of sugar to starch. This type is our modern sweet corn. The kernels on the cross-bred ears that Shull harvested at the end of the summer were all round, starchy kernels. Two versions of the same gene determine whether the kernels are wrinkled and sugary or full and starchy. (The word "gene," in this modern meaning, would not be used until 1909; Shull still called them hereditary factors.) The starch-making version, or allele, was dominant, while the mutant sugar-making allele was recessive. If pollen from a starchy plant is brushed onto the silks of a sugary plant, each resulting kernel receives a functional starchy allele from the pollen parent and a non-functioning sugary allele from the ear-bearing parent. Such a kernel would be heterozygous (from the Greek words for "different" and "yolk").

When Shull took these heterozygous kernels, grew them, and crossed them with a plant that had the same sugary allele from both parents (and so was *homozygous*, from the Greek for "the same" and "yolk"), half of the kernels on each ear were round and half were wrinkled. The traits, each determined by a different form of a single gene, had segregated from each other and the wrinkled trait had reappeared. The experiment thus provided the demonstration of Mendelian inheritance that Davenport had requested.

But Shull also noticed that the number of rows of kernels on the two types of ears—starchy or sugary—was quite different. Curious about how this second trait was inherited, he cross-pollinated plants grown from the different parent ears and counted the number of rows on the resulting ears. He also self-pollinated the plants. This self-pollination, or inbreeding, Shull saw, was not good for the plants. The progeny of self-pollinated plants were smaller and more disease-prone and had smaller ears. Yet when he crossed these puny inbreds and grew the kernels, he was startled to get much healthier plants that produced much bigger ears.

He published these results in 1908. A year later he explained how his observations could be put to practical use in a short paper called, "A pure-line method of corn breeding." He admitted that this method of producing seed-corn was probably more expensive, and he was unsure if the increase in yield would be enough to cover the cost. "These are practical questions which lie wholly outside my own field of experimentation," he concluded, "but I am hoping that the Agricultural Experiment Stations in the corn-belt will undertake some experiments."

The Agricultural Experiment Stations were not thrilled. Shull's method was "impractical," "too complex," and "not cost-competitive," being three times more expensive than open-pollinated corn. Producing enough seeds of the inbred lines would be a bottleneck, making it harder for a farmer to get seed. Farmers were accustomed to saving corn from their own crops for planting the next year. They did buy seed corn occasionally to freshen their own supply, but with the new hybrid varieties they would have to buy new seeds every year. Moreover, the technique, wrote one USDA breeder, echoing Johnny Appleseed and in turn to be echoed by Prince Charles and Congressman Kucinich, did "violence to the nature of the plant." It was "dangerous." Those experiment stations that did try the technique, writes Charles Heiser in *Seed to Civilization*, thought the inbreds were "such sick-looking plants" that they grew them only "in out-of-the-way places where farmers would be unlikely to see them." Otherwise, "they would think the breeders were working in the wrong direction."

Yield, in fact, was not the standard by which corn was judged in the early 1900s; beauty was. At the corn shows popular throughout the country, a farmer would enter a single perfect ear, hoping to win an expensive tool or machine. At the National Corn Exposition each year, one ear would be named Champion of the World. The judges, many of whom were university scientists, would rank the ears on aesthetic factors, even though they knew that beauty had little to do with performance. A prize-winning ear, according to one show card, was "10 $1/_2$ inches long, 7 $1/_2$ inches in circumference, with 20 or 22 straight rows of kernels carried out to the tip, and with a well-rounded butt. The kernels had to be moderately wide, keystone in shape, deep, plump at the tip, and without any trace of being shrunken or blistered." Some judges (especially in Iowa and Illinois) gave extra points if the kernels were "horny and shiny." By 1910 it was clear that the characteristics weighted heavily on the show card actually lowered yield, but the shows were so popular among farmers that change came hard.

The breakthrough occurred in 1917, the same year that the Hungarian agricultural engineer (and owner of the largest pig farm in Europe) Karl Ereky coined the term "biotechnologie." It was also the year of the Russian Revolution (with its slogan "Bread and Peace") and the Plow to the Fence for National Defense campaign in support of World War I. In that year Donald F. Jones produced a Mendelian explanation for Shull's results.

Jones was a student of Edmund M. East, who earlier, while at the Connecticut Agricultural Experimental Station, had pronounced Shull's hybrid corn idea "impractical," "too complex," and "not cost-competitive." Jones's double-cross hybrid method, published in 1918, solved the practical, economic problem (though not the complexity). Rather than Shull's two inbred parent lines, Jones used four. The plants that produced the seeds farmers would buy were not the puny, sickly inbred lines (cause of the bottleneck East predicted), but more productive single crosses. A few bushels of single-crossed seed would produce a thousand bushels of double-crossed seed which, when planted by farmers, would show the increased vigor Shull discovered, now called heterosis.

Yet Jones's double-cross hybrids still had no chance of winning at a corn show: the ears were far from the show ideal. Farmers were also not convinced that buying seed every year would benefit anyone other than the seed dealer. Henry A. Wallace, an editor, with his father, Henry C. Wallace, of *Wallace's Farmer,* came up with the solution in 1919. In the pages of the magazine, he suggested that the corn show branch out into a new contest: a yield test under controlled conditions. Iowa State University picked up on the idea and in 1920 started the Iowa corn yield test. Four years later Wallace won the gold medal with Copper Cross, the progeny of two inbred lines, one of which came from Jones. In 1926 Wallace founded the Pioneer Hi-Bred Corn Company, marking the transition of the seed dealer's market from backyard gardeners to farmers, who previously had supplied almost all of their own seeds, saving some from each harvest.

In 1933 only 1 percent of the corn grown in the United States was hybrid corn. The average yield was 23 bushels per acre. Then came a drought that destroyed most traditional varieties. Plant breeder Don Duvick, who recently retired from Wallace's Pioneer Hi-Bred company as Vice President for Research, recalls from his childhood:

> We planted our first hybrid corn in 1936, a year of disastrous drought throughout the Corn Belt, including our section of northeastern Illinois. It was on only a small part of our corn acres; most of them were planted to our "tried and true" open pollinated variety. The only grain Dad harvested was from the hybrid; the rest wasn't worth trying to harvest for grain; he cut the barren stalks and brought them in for badly needed forage for the cattle. From then on, we planted only hybrids. But in 1937, Dad did plant a

small area to some F2 seed of the hybrid from the previous year. He had been told that yield would drop off, but wondering if the "professors" were right, he decided to make his own check. They were right; the F2 looked terrible (I can remember that myself). He grinned and said, "Well, I just wanted to see for myself."

By 1940 30 percent of U.S. corn was hybrid; the average yield had risen slightly, to about 30 bushels per acre. By 1970, when Ingo Potrykus began his genetic studies of petunias, 96 percent of U.S. corn was hybrid. The average yield was a remarkable 72 bushels per acre.

Though he was most likely not in the audience when Luther Burbank lectured in 1925 at the First Congregational Church in San Francisco, Henry Wallace might have agreed when Burbank proclaimed, "What a joy life is when you have made a close working partnership with Nature, helping her to produce for the benefit of mankind new forms, colors, and perfumes in flowers which were never known before; fruits in form, size, color, and flavor never before seen on this globe; and grains of enormously increased productiveness, whose fat kernels are filled with more and better nourishment, a veritable storehouse of perfect food—new food for all the world's untold millions for all time to come."

The year Wallace founded his Pioneer Hi-Bred Corn Company, the world population was 2 billion. By 1950, when the first mutation-bred crops formed by irradiating seeds were being sown, and F. G. O'Mara was using colchicine to develop the fertile wheat-rye hybrid triticale, the world's population had grown to 3 billion. Most of the fertile land on Earth was already in production by then and the clearing of new agricultural land would be balanced, over the next 50 years, by loss of fertile land to urbanization, desertification, and salinization. Meanwhile, more than 70 million human beings were added to Earth's population each year. Pundits were predicting mass starvations: up to a billion deaths.

Wallace had much to do with the fact that this food crisis was averted. In 1943, having served as U.S. Secretary of Agriculture from 1933 to 1940, he was the Vice President of the United States under

Roosevelt. He tried to persuade Congress to provide agricultural aid to Mexico. Having failed, he turned to the Rockefeller Foundation, the philanthropic organization pledged to "applying science to benefit mankind." An agreement was made with the Mexican government to support a new kind of cooperative technical assistance program. Its first objective would be to improve local crops, including corn, wheat, potatoes, and beans. Mexican students would be trained in agricultural science and helped to set up their own research programs.

Norman Borlaug had just completed a doctorate working on rust, a fungus that plagues wheat and other crops, and was looking for a job. In 1944 he became director of the wheat program, "a job for which there was little competition, backwater Mexico in the 1940s not being an eagerly sought-after posting," notes one writer. In his 1970 Nobel Peace Prize lecture, Borlaug recalled, "At that time, Mexico was importing more than 50 percent of the wheat that it consumed, as well as a considerable percentage of its maize. Wheat yields were low and static, with a national average yield of 750 kilos per hectare, even though most of the wheat was grown on irrigated land." The soils were "impoverished," and chemical fertilizer "virtually unknown."

Research at the Mexican institute "from the outset," said Borlaug, "was production-oriented and restricted to that which was relevant to increasing wheat production. Researches in pursuit of irrelevant academic butterflies were discouraged. As soon as significant improvements were made by research, whether in varieties, fertilizer recommendations, or cultural practices, they were taken to farms and incorporated into the production programs. We never waited for perfection in varieties or methods, but used the best available each year and modified them as further improvement came to hand." Borlaug and the other researchers demonstrated the new techniques on the farms themselves. "This forced the research scientists themselves to consider the obstacles to production that confronted the farmers."

When World War II ended and nitrogen-based explosives were no longer needed, the price of nitrogen fertilizers fell and their use increased. Plants grew bigger and yields climbed, but soon a new problem arose: lodging. The tall stalks, topped by heavy heads of grain, fell over in wind or rain and could not be harvested. By 1953 Borlaug was searching for a small, sturdy wheat variety that the wind and rain wouldn't knock down.

A few years before, a USDA scientist, S. C. Salmon, had brought seeds of Norin wheat, a group of dwarf varieties, home from Japan. As a visitor to Japan had remarked back in 1873, "The Japanese have made the dwarfing of wheat an art. The wheat stalk seldom grows longer than 50 to 60 centimeters. The head is short but heavy. No matter how much manure is used, the plant will not grow taller; rather the length of the wheat head is increased. Even on the richest soils, the wheat plants never fall down." Salmon shared the seeds with several breeders, including a research group at Washington State University who bred a hybrid strain habituated to the Pacific Northwest. It produced a world record of 216 bushels per acre (over 14,000 kilos per hectare). The Washington group sent seeds of their champion hybrid, as well as some of the original Norin seeds, to Borlaug. His first crop of the dwarf wheat, planted near Mexico City, was lost to rust—the very fungus he had studied for his Ph.D. research. The next year he planted near Mexico's west coast, where the wheat did well. "Next he crossed his new dwarf strain with everything else he had around," according to the authors of *Biology: A Human Concern*, "and this was the beginning of the revolution, so to speak." The revolution was the Green Revolution.

As Borlaug explained in his Nobel lecture, "Through a series of crosses and re-crosses begun in 1954, dwarfness was incorporated into the superior, new-combination Mexican types, finally giving rise to a group of so-called dwarf Mexican wheat varieties." By changing the plant's architecture to emphasize a short, sturdy stalk, the dwarfness trait allowed the wheat to produce heavier seed heads—given enough water and nitrogen—without falling over in a breeze. In addition, the plants were not affected by length of day (and so could grow at a range of latitudes) and were highly resistant to wheat rusts. The result, in Borlaug's terms, was a "yield blast-off." A few seasons after the new variety was introduced Mexico became self-sufficient in wheat. When introduced into Pakistan and India, the wheat had the same yield-boosting effects.

At the same time as Borlaug was searching for dwarf wheats, re-searchers at the new Ford- and Rockefeller-funded International Rice

Research Institute (IRRI) in the Philippines were looking for dwarf rices—what came to be known as "miracle rices." In 1963 Hank Beachell, a retired USDA scientist, found a short, thick-stemmed rice plant in row 288 of an IRRI experimental field. It had come from a cross between a short, stiff-strawed variety called Dee-geo-woo-gen from Taiwan and a taller, vigorous, pest-resistant variety called Peta from Indonesia. It was the eighth cross made, so Beachell named it IR8.

Combining the vigor of Peta with the shortness of Dee-geo-woo-gen, IR8 grew more branches, or tillers, each of which produced grain, giving each plant a higher yield. Like Borlaug's wheat, IR8 was insensitive to day length; it could mature 60 days sooner than other varieties, allowing farmers to grow two crops a year instead of one.

The genetic change behind this miracle rice was not discovered until 2002. Ashikara Motoyui and Matsuoka Makoto of Nagoya University in Japan reported that they had isolated the gene (called *semidwarf 1* or *sd1*) that was responsible. This gene codes for a protein needed to produce a plant hormone. The *sd1* allele contained a large hole, a deletion: it was missing 383 pairs of nucleotides, or base pairs, out of a total of 1,170. This mutation made the gene useless—with extraordinary consequences. Nitrogen fertilizer normally increases both the amount of grain a rice plant produces and its height, the same way it does in wheat. Tall rice, like tall wheat, is vulnerable to lodging, being knocked down by wind and rain. The IR8 plants, however, stayed short when fertilized and they stayed upright until they were harvested. Their sturdiness in itself increased yield. But there was an added bonus. Because the plant did not expend as much of the energy it harvested from the sun in growing stems and leaves, more went into making grain. The fraction of a plant's total weight that is in the grain is termed its harvest index. The harvest index of IR8 and the other new semi-dwarf rice varieties increased from about 30 percent to more than 50 percent.

When IR8 was released to farmers in 1966 it changed Asian agriculture. IR8 was rapidly adopted in the Philippines, then India, Pakistan, and Indonesia. A year later Indian scientist Gurdev Singh Khush joined the IRRI staff. Khush had a hand in breeding more than 300 rice varieties, including IR36, which, according to IRRI, was "the most widely planted variety of rice, or of any other food crop, the world has ever known." With the new varieties, rice yields more than doubled from the mid-1960s to 1990 throughout Asia. Upon Khush's retirement in 2001 the IRRI annual report claimed that "in any rice field, anywhere in the world, there's a 60 percent chance that the rice was either bred at IRRI under his leadership or developed from IRRI varieties." Said Norman Borlaug, "The impact of Dr. Khush's work upon the lives of the world's poorest people is incalculable."

It was to Khush's program at IRRI that Karabi and Swapan Datta came, straight from Ingo Potrykus's laboratory, in 1993. And it was here that Golden Rice was sent in 2001, to be grown in a high-security, Biosafety Level Four greenhouse.

4 GENES AND SPECIES

The classification of nearly all cultivated plants is now in such a perilous state that the use of any such Latin term as Phaseolus vulgaris *is just an elaborate and technical way of saying, "I do not know."*

—Edgar Anderson (1967)

Teosinte is no longer called *Euchleana mexicana.* Once it was accepted as the ancestor of corn, its genus name, *Euchleana,* and corn's genus name, *Zea,* were lumped into one classification and teosinte became known as a subspecies, called *parviglumis,* of corn, *Zea mays.*

Likewise when the goat grass called *Aegilops squarrosa* was found to have contributed its entire genome to bread wheat, it too lost its name and species classification. Rather than following Edgar Anderson's whimsical suggestion that the wheat genus be renamed *Aegilotriticopyron,* to give due credit to its roots, taxonomists decided to demote *Aegilops squarrosa* and the other 20 or so *Aegilops* species and put them into the wheat genus, *Triticum. Aegilops squarrosa* was duly renamed *Triticum aegilops.* Other taxonomists, however, refused to use the new name (or preferred yet a third name), and international symposia have been convened several times to debate the value of the switch. "The name issue is not yet settled," argued L. A. Morrison of Oregon State University in 1998. "There is confusion about the number of species in the genus *Triticum.* There is confusion about the correct names of the species." Morrison urged other wheat researchers to "return to the traditional concepts of *Triticum* and *Aegilops* as separate genera."

First teosinte is a species in a different genus, then it is a subspecies of corn. First *Aegilops squarrosa* is a species in its own distinct genus, separate from wheat, then it is not and its entire genus disappears. Though it might reappear. The plants in that genus are not extinct, of course. Indeed, they have not changed at all—only our way of looking at them has changed. The classifications "genus" and "species" are not fixed and immutable. They are, explains Ernst Mayr, author of *Systematics and the Origin of Species*, conveniences: they provide "a reliable, easy-to-use, filing system for the bewildering diversity of nature." C. F. Bessey noted in the *American Naturalist* in 1908, "Nature produces individuals and nothing more. Species have been invented in order that we may refer to great numbers of individuals collectively."

The filing of plants into a system of genus and species was begun in 1753 by the Swedish botanist Carl Linnaeus, who was also the first to prove that plants were sexual creatures like animals. The tradition of giving each plant a two-part name (first genus, then species) has held since then, but the rules have changed. Linnaeus, like Aristotle long before him, classified plants and animals chiefly by how they looked. Paleontologists use a similar idea of species to classify fossils. Their "morphospecies" are often based on single fossils. Yet individuals in the same species can differ considerably in size. Should each size of dog be a new species? This definition also fails to account for organisms with very different-looking life stages. Should the caterpillars and the butterflies they turn into belong to different species?

Seeking to include evolution in the definition of species, paleontologist George Gaylord Simpson called a species "a series of ancestor descendent populations passing through time and space independent of other populations, each of which possesses its own evolutionary tendencies and historical fate." The problem with Simpson's definition is that there are gaps in fossil records, and such gaps can create arbitrary boundaries between species.

Searching for a better definition of "species," we simply find more definitions. Ernst Mayr's biological concept, that "Species are groups of actually or potentially interbreeding populations which are reproductively isolated from other such groups," is similar to one of the earliest definitions, that of John Ray, who in 1682 defined a species as "a set of individuals who give rise through reproduction to new individuals similar to themselves." Population biologist Theodosius Dobzhansky early in the twentieth century said that species "are systems of populations; the gene exchange between these systems is limited or prevented in nature by a reproductive isolating mechanism or several such mechanisms." Yet the mechanisms Dobzhansky had in mind include being on the other side of a mountain, or living in the top of a tree rather than at the bottom. By this definition, people living on different continents in the days before ship travel belonged to different species.

Defining a species as a group of individuals that look alike or can breed together is even more problematic in plants. Indica rice and Japonica rice, for example, are two popular types of the cultivated rice, *Oryza sativa.* They are members of the same species, and it is often difficult to tell if a single grain comes from one type or the other. While Japonica is usually short-grained and Indica is long-grained, more than a third of the time a grain of Indica is short enough to look like Japonica and a grain of Japonica is long enough to seem to be Indica. Likewise the two types overlap in how they respond to drought, cold, pesticides, and other challenges. Yet they do not crossbreed. When scientists force the crossing—using laboratory techniques very much like those called into play to help an infertile woman bear a child—the resulting Indica-Japonica hybrids are sterile and do not set seed. The two types of rice have been separate for at least 7,000 years. Yet they are still considered the same species.

Then, on the other hand, there are the brassicas. In the grocery store, cabbages, kale, kohlrabi, cauliflower, broccoli, and Brussels sprouts are all neatly divided into their own bins. Diners who ordered Brussels sprouts in a restaurant would be outraged if they were served cabbage instead. And yet all of these common vegetables—even the purple Graffiti cauliflower—are members of the same species, *Brassica oleracea.*

The original cabbage was apparently native to the Mediterranean coast, where it was grown for its oily seeds, although wild cabbages also

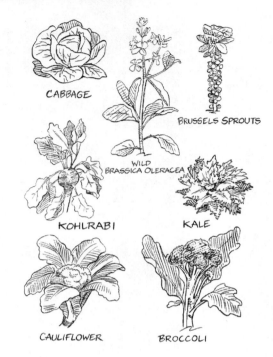

CABBAGE

BRUSSELS SPROUTS

WILD BRASSICA OLERACEA

KOHLRABI

KALE

CAULIFLOWER

BROCCOLI

The many faces of Brassica oleracea

grow along the northwestern coasts of Europe. The Greeks were growing leafy kales, probably for livestock feed, by 2,500 years ago. The Romans enjoyed loose-headed cabbages. True hearting cabbages first appeared in Germany, in both white and red varieties, in the 1100s. Cauliflower, kept white by tying its leaves over its head, was being grown in Europe by the 1500s. The advent of kohlrabi, with its swelled stem, is unrecorded, but broccoli, then called "Italian asparagus" or "sprout colli-flower," began to be popular in the late 1600s. Perhaps people had become more patient by then. As one food historian notes, "I am not certain people realize that many broccolis are at least a year old when they eat them. Like other cabbages, broccolis usually require two years to come to flower, and it is the flower that we are eating." If allowed to open, he noted, "the flowers would be waxy yellow and fragrant. Butterflies adore them." Last to the table came Brussels sprouts, showing up as a sport (or mutant) in a Belgian garden in 1750.

If grown carefully, with no chance of sharing pollen, each of these varieties of *Brassica oleracea*—as well as all their subtypes, such as purple cauliflower—breeds true: broccoli seeds produce more broccolis, cauliflower seeds more cauliflowers, and so on. But chaos awaits the unwary gardener. Brassicas are outbreeders. They prefer—they are "anxious," say the botanists—to mate with plants that are genetically unlike themselves. Some are even self-incompatible, meaning that they will not use their own pollen to fertilize their own flowers even if no other pollen is available. (Apple trees are self-incompatible too, which is why orchardists plant different varieties in alternating rows.) In an experiment in England, botanist W. F. Giles planted a mix of garden-variety cabbages and kales in a square and allowed them to cross-pollinate. "The work of centuries was wiped out in a generation," reports one observer. The cabbage heads disappeared and the wide, branching, gangly look of the wild plant took over.

And the promiscuous *Brassica oleracea* tribe will not only mate with each other. Some variety of cabbage in seventeenth-century Bohemia crossed with a turnip, which is in the genus *Brassica,* but a different species. This cabbage-turnip cross is the rutabaga. Rutabaga is classified as a variety of *Brassica napus.* Another variety of that species is now widely grown—like the original domesticated cabbage—for its oily seeds, the source of the popular canola oil.

〜〜〜

The danger of crossing the species barrier—even the immorality of it—is often cited as a reason to restrict, or even prohibit, the molecular modification of plants. In 1984 activist and writer Jeremy Rifkin testified before Congress that transferring a gene from one species to another represents "a fundamental assault on the principle of species integrity." In 1998 the authors of *Genetically Engineered Foods: Are they safe? You decide* wrote that "The mixing of genes by cross-breeding is clearly subject to very definite rules—you can't mix unrelated species. . . . Where there are rules there are boundaries. . . . Natural law has set a boundary. Genetic engineering is not constrained by these rules and crosses all boundaries set in place by natural law." To support their case, they quote John Hagelin, a physicist: "When genetic engineers disregard the reproductive boundaries set in place by natural law, they run the risk of destroying our genetic encyclopedia, compromising the richness of our natural biodiversity and creating 'genetic soup.' What this means to the future of our ecosystem, no one knows."

In 2000 Representative Kucinich argued in Congress that "Conventional breeders are bound by species boundaries that allow them to transfer genetic material only between related or closely related species. By contrast, the very purpose of genetic engineering is to allow scientists to transfer genes from completely unrelated life forms." In 2001 journalist Bill Lambrecht defined "genetically modified organism" in *Dinner at the New Gene Café* as "what you get when you move genes across the traditional species boundaries of plants and animals in the quest for new traits."

These critics are apparently operating under yet another understanding of the word "species." One such definition comes from the Middle Ages, when scholars rediscovered the works of Plato and Aristotle and reinterpreted them in a way that had a lasting effect. Johnny Appleseed's Swedenborgian notion that every apple tree was a reflection of a spiritual tree, and therefore should not be grafted or otherwise kept from expressing itself, has its source in this medieval worldview. According to Thomas Aquinas, the thirteenth-century Italian philosopher, everything on Earth is but a reflection (albeit a poor reflection) of what exists in Heaven. Each species represents an idea in the mind of God.

This medieval concept of species was a major stumbling block for the theory of evolution when it was first introduced in the mid-nineteenth century. The French scientist Jean-Baptiste Lamarck set forth in 1809 the idea that creatures had changed—or evolved—since God created Earth. This idea seemed to explain satisfactorily how a cabbage could become a broccoli. But Darwin, in *The Origin of Species*, went further. He argued that one primordial plant could branch into many lineages, giving rise to all flowering plants, from broccoli to apple to daffodil. (Not to mention that one primordial ape could give rise to both monkeys and humans.)

Even scientists who did not think Darwin's argument was blasphemous thought they saw a flaw in his reasoning. Darwin's theory of natural selection assumes that individuals within a species differ in how fit they are. Those whose fitness best matches the challenges of their home environment survive to have more children and pass down to those children something that will also, by and large, make them more fit. The accumulation of small improvements in fitness, generation by generation, eventually leads to the branching off of a new and better species.

The problem lay in that first assumption. Everyone agreed that individuals were different. But were those differences inherited? On what did natural selection act? "Not until the 1930s was this missing element filled in," notes Harvard physiologist Bernard D. Davis. That element was the gene. To know what species—or species barrier—means to a modern-day plant scientist, to see what Golden Rice, triticale, bread wheat, broccoli, hybrid corn, and canola oil might have in common, takes an understanding of the gene.

Like the notion of a species, the concept of the gene went through many revisions before a scientific consensus was reached and, like species, it has changed dramatically since then, as we learned more about what actually occurs inside a cell.

The field of genetics, the science of heredity, began in 1900 when Hugo De Vries and Carl Correns rediscovered the pea experiments Mendel had done in his monastery garden half a century earlier. Correns wrote, "The same thing happened to me which now seems to be happening to De Vries: I thought that I had found *something new*. But then I convinced myself that the Abbot Gregor Mendel in Brünn, had, during the sixties, not only obtained the same result through extensive experiments with peas, which lasted for many years, as did De Vries and I, but had also given exactly the same explanation, as far as that was possible in 1866."

In Darwin's and Mendel's time, the mechanics of inheritance were not well understood. Most people in the nineteenth century thought of it as a blending of traits, like a mixing of water and wine or two colors of paint. What Mendel found was far from blending and so did not seem to relate to ordinary inheritance. Mendel crossed pea plants that bore round seeds with pea plants that bore wrinkled seeds. When he planted the seeds of the cross-pollinated plants, the next generation (called the first filial, or F1, generation) bore only round seeds. The seeds were not only less wrinkled than their wrinkly parent (as they might have been if the two traits had blended), they had no wrinkles at all. The trait had disappeared. But curiously, when these F1 plants were allowed to pollinate themselves (which peas prefer to do, being natural selfers, not anxious outbreeders like the brassicas), they produced round and wrinkled seeds in a ratio of three round to one wrinkled. The trait that had vanished in the F1 generation re-emerged unchanged in the F2 generation.

Mendel showed that this behavior was simple, regular, and reproducible. His explanation was simple, elegant—and revolutionary. Traits, or as he called them, "differentiating characters," underwent no blending at all. Each organism had two hereditary units for each trait, one from the father and one from the mother. Only one of these two units would determine the trait. This one was dominant. The other, the one that vanished in the F1 generation, he called recessive. If a seed that received both a dominant (roundness) unit and a recessive (wrinkliness) unit was round in appearance, then the three-to-one ratio was easily explained. Because half of the parents' hereditary units called for round seeds and half called for wrinkled seeds, one quarter (half of a half) of the offspring would receive only roundness units and another quarter would receive only wrinkliness units. The rest would receive a roundness unit from one parent and a wrinkliness unit from the other parent. Because roundness obscures wrinkliness, three-quarters of the seeds would be round and only one-quarter—that quarter receiving two wrinkliness units—would be wrinkled.

These hereditary units did not change even when they could not be seen. Nor did the inheritance of one trait seem to affect the inheritance of another. When Mendel bred not only round and wrinkled peas, but green and yellow ones, he found that the two traits "assorted" independently. Only 1 in 16 (a quarter of a quarter) of the F2 offspring were both wrinkly and yellow, meaning that they had two recessive units for each trait. But while Mendel could predict how many of the plants in his garden each season would bear yellow peas, he could not explain how the process worked.

As the field of genetics began to develop in the twentieth century, first the units of heredity were named, then they began to acquire a physical identity. De Vries, who had coined "mutation," decided to call

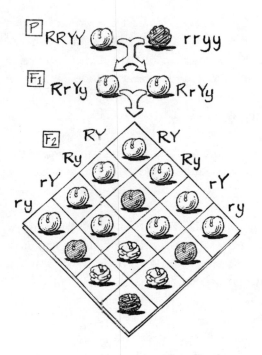

Mendel s experiment with peas, showing the assortment of round (R), wrinkly (r), yellow (Y), and green (y) traits

them pangenes, from Darwin's concept of pangenesis, which said that every cell in the organism sent a tiny bit of itself to the egg or sperm so that the resulting embryo would know how to recreate a whole organism. Other geneticists came up with the names plasomes, idioblasts, and biophores for the same particles. Each name had an earlier theory attached to it, and none of them satisfied everyone. Then in 1909 Danish botanist Wilhelm Johannsen published his work on beans. "It appears simplest to isolate the last syllable, 'gene,'" he wrote, from De Vries's pangene. "The word 'gene,'" he noted, "is completely free from any hypotheses." The term stuck, as did Johannsen's definitions of phenotype, for the outward appearance or character of an organism (in Mendel's peas, wrinkled or round), and genotype, for the combination of gene variants (or alleles) the organism carried: either two dominant types, two recessive types, or one of each.

Still, it took a while before scientists agreed that these particles, these genes, were things—that they had physical form that could be observed and changed. In 1912 Edmund M. East, the expert on corn genetics who thought hybrid corn was a waste of effort and believed teosinte could not possibly be the ancestor of corn, thought the gene was only a useful shorthand. "As I understand Mendelism, it is a concept pure and simple," he wrote, "a conceptual notation as is used in algebra or chemistry. No one objects to expressing a circle as $x^2 + y^2 = r^2$."

Two years later Thomas Hunt Morgan at Columbia University would claim that a gene was "some minute particle of a chromosome whose presence in the cell influences the physiological processes that go on in the cell." Chromosomes had long been observed in the cells of both plants and animals. In 1875 the German botanist E. Strasburger had described spindly structures that, as the cell divides, first lengthen, then shorten again to form compact little rods. In 1888 botanist Wilhelm Waldeyer noticed that these rods took up a stain well, so he named them "chromo" (colored) "somes" (bodies). But no one had, as yet, connected them to genes.

Morgan, who would do so, began graduate school at Johns Hopkins University in 1886. There he was taught by a self-proclaimed Darwinist, William Keith Brooks, who had worked with William Bateson, the geneticist who later introduced Mendel's laws to English science. Brooks tried to interest Morgan, as he did all of his students, in the study of heredity. But after completing a dissertation that used a comparison of embryos to prove that sea spiders were true spiders, not crustaceans like lobsters, Morgan went to Italy to continue studying how eggs develop.

One theory at the time suggested that within each egg was a tiny manikin, a miniature creature or homunculus, that could somehow unfold into an adult. To test this theory, Morgan took a two-celled embryo of a whitefish and destroyed one cell. The remaining cell grew into an "incomplete" adult, he found—unless he turned it upside down, shook it, or whirled it in a centrifuge. Then it developed normally. Clearly things were more complicated than the manikin theory supposed.

In 1903 Morgan went to Columbia University as America's first professor of experimental zoology. Inspired by Mendel's laws of heredity and De Vries's mutation theory, he experimented on all sorts of animals. He used cockroaches, crabs, tadpoles, sea urchins, starfish, earthworms, flatworms, jellyfish, mice—by one estimate, more than 50 different creatures—before he settled on the fruit fly, *Drosophila*. And unlike his predecessors, who had contented themselves with observing the bewildering variety of nature, Morgan and his students induced mutations, creating odd-looking flies (yellow bodies, white eyes, pink eyes, tiny wings) by exposing them to heat, cold, salt, sugars, acids, alkalis, radium, and X-rays.

His laboratory was known as the Fly Room. It "was a rather small room," recalled Alfred Sturtevant, "with eight desks crowded into it, in which the three of us reared *Drosophila* for the next 17 years." Sturtevant joined the Fly Room in 1909. An undergraduate, a teenager, he had impressed Morgan with a paper on the inheritance of coat color in his family's stable of racehorses. He, in his turn, was impressed by a lecture in which Morgan described what he called the strength of coupling or linkage of certain genes. Strangely, these fruit fly genes did not assort independently, as had the genes of Mendel's peas. Some appeared to stick together—to be coupled to each other.

In Mendel's peas, the fact that an offspring received one trait (for instance, wrinkly seeds) made it no more or less likely that it would receive a second trait (green seeds). But in Morgan's fruit flies, some traits tended to be inherited together. If a fruit fly offspring had white eyes, it was also likely to have a yellow body. The reason, Morgan believed, had to do with where the genes physically resided in the cell. Genes, he argued in his book *The Mechanism of Mendelian Heredity*, were not abstract concepts. They were physical units, located at definite positions on the chromosomes. Genes that were linked must be near each other on the same chromosome. The closer they were, the more likely the two traits would be inherited together. The farther apart they were, the more likely it was that they would not be inherited together. The reason has to do with the way chromosomes exchange parts or recombine during meiosis, when the sex cells were formed.

It was Morgan who named such recombination "crossing over." Crossing over happens just before the chromosomes split into two haploid, or half sets, each of which furnishes one normal egg or sperm with its genetic material. Each DNA helix in the chromosome is cut in several places and a bit is chewed off each end. The loose ends invade their counterpart chromosome, the homolog (from the Greek for "agreeing in proportion") received from the other parent, pushing out one strand of its DNA. The homologs exchange parts, each carrying genes from the previous generation. Then the DNA strands in the chromosomes are reconnected. Some genes received from the grandmother are now attached to genes received from the grandfather. In this way, genes from grandmother and grandfather are mixed before being passed down to the grandchild. Genes that are linked are located on the same cut bit of chromosome, and so are passed down together.

Sturtevant was intrigued by Morgan's idea. He talked his way into Morgan's lab and devised a test for his hypothesis. Like Morgan, Sturtevant and another student, C. B. Bridges, kept their mutant flies in hundreds of half-pint milk bottles, feeding them mashed bananas. The milk bottles were hardly a high biosafety-level containment facility for these genetic experiments. "Near the entrance of the room a stalk of bananas hung conspicuously, serving as a center of attraction for the numerous fruit flies that had escaped from their milk bottles or that had bred themselves, without the benefit and direction of science, in the garbage can that was never thoroughly cleaned," wrote Ian Shine and Sylvia Wrobel in their biography of Morgan. The lab window was often left open, both to lure in flies that could then become breeding subjects and because the room stank of fermented banana. Morgan had other experiments going on in the lab, including some "whose purpose no one ever figured out exactly, like the one in which a crab walked around with another crab glued to its back, a fragment of radium between the pair."

By tracking fruit fly mutations over generations, Sturtevant was able to draw the first genetic map. Published in 1913, it showed six genes—those for yellow bodies, white eyes, vermilion eyes, miniature wings, and rudimentary wings—in a straight line on the fruit fly's sex chromosome. Morgan's definition of a gene seemed correct. A gene was a specific, fixed point on a chromosome: a locus, from the Latin for "place." Genes were like beads on a string, like pearls in a necklace, like dots on a line. The gene had been defined as the unit of heredity.

It soon became obvious that the number of linkage groups in a fruit fly, a pea, or a person was the same as the number of chromosomes in a cell. Knowing this, scientists were able to tie each linked group of genes to a chromosome. By 1936 corn breeders, for instance, could make use of a map that located 100 corn genes onto specific chromosomes. They could track the pattern of inheritance of another 250 genes from crop to crop.

Each bead on the string, each gene, directed the production of one enzyme, George Beadle and Edward Tatum at Caltech determined in 1941 after studying mutations in orange breadmold. (Later Linus Pauling, at the same institution, would expand their idea: each gene controlled production of one *protein*, he said, of which enzymes are only one kind). In 1944 Oswald Avery, Colin MacLeod, and Maclyn McCarty at the Rockefeller Institute for Medical Research in New York carried out experiments that led to the identification of the chemical nature of genes. They found that a chemical called deoxyribonucleic acid, or DNA, extracted from a virulent bacterium, could by itself transform a harmless form of pneumonia bacteria into a virulent type when the two were injected into the same mouse.

Until then DNA had been dismissed as a "stupid" molecule, in the words of physicist Max Delbrück. It was thought to be too simple to be of much biological importance. It was made up of just four small molecules, the nucleotides, each of which had a sugar, a nitrogenous base, and a phosphate group. Nor were the bases themselves of much interest. Adenine (now known familiarly as A) was common in beef pancreas. Cytosine and thymine (C and T) were purified from the thymus glands of calves. Guanine (G) had first been found in guano (bird feces); in crystallized form, it gave the shine to fish scales.

Proteins, on the other hand, had long been thought special. Discovered in the 1830s, they had been named from the Greek *proteios*, meaning "of the first importance." They were long chains, or polymers made up of 20 different small molecules, the amino acids. Proteins might be complex enough to make a gene, but certainly DNA was not. Then in 1952 Al Hershey and Martha Chase showed that when a virus called a bacteriophage infects a bacterial cell, it injects only its DNA into the cell, abandoning its protein coat. Once again, the active ingredient was DNA. Suddenly DNA was the "molecule of life."

How could complex genetic information—the making of a white eye or a wrinkled pea—be conveyed by a simple molecule with only four constituents, A, T, G, and C? The secret lies in how those parts are assembled—both in the order of the letters and in the structure of the DNA molecule. As James Watson and Francis Crick announced in a short paper in *Nature* in 1953, DNA is a double helix, rather like a spiral staircase. Watson and Crick had earlier proposed to their London colleagues, Rosalind Franklin and Maurice Wilkins, that DNA was a triple helix. Franklin, however, pointed out that they had mistaken (by a factor of 10) the proportion of water in the structure. Franklin was a crystallographer. She took photos of the diffraction patterns created when crystals of DNA were exposed to X-rays. Studying the patterns, she was learning about the molecule's structure. She had not yet published her photos, or fully interpreted what they meant, when Wilkins arranged for Watson to look at the best of them. In 1962 Watson, Crick, and Wilkins received the Nobel Prize in physiology or medicine for discovering the double helix. (Franklin died of cancer before the prize was awarded.)

"The novel feature of the structure," Watson and Crick wrote of the double helix, "is the manner in which the two chains are held together." Because A can pair only with T, and G can pair only with C, the two chains are mirror images of each other. "It has not escaped our notice," Watson and Crick wrote in the paper's final statement—or understatement—"that the specific pairing we have postulated immediately suggests a possible copying mechanism for the genetic material."

By 1969 the DNA code was declared cracked. Each sequence of three letters, each codon, such as GAG or GAA or AAA, stands for an amino acid (except for three codons, which instead mean "stop"). When the code is read, the cell's machinery links the specified amino acids together into a chain, which is then folded up to become a functioning protein. So each protein made by the body has its corresponding DNA code—a gene. (Scientists generally say the gene encodes, or codes for, the protein.) Each bead on the string, each gene, turned out to be more like a sentence written in three-letter words. Each sentence was the recipe for a protein. Mendel's "differentiating characters" could thus be explained as alternative forms of the same gene: one codes for a normal protein; the other does not because one or more words in the sentence had changed, or mutated.

The genome—the sum total of an organism's DNA—was understood to be its book of life, life's little instruction book. The analogy is, of course, not exact. A true book contains many pages of sentences folded in upon each other: you read across a line of letters, then go to the beginning of the next line. To turn a true book into the book of life, you must take out the spaces between words. Connect the last letter on each line with the first letter of the first word on the next line until you have strings of letters folded one upon the next. Then grab the last word in the book and pull—the letters of the book will stretch on for miles. So it is with DNA.

This long string of meaning, made up of words and punctuation all fashioned from the same four letters, A, T, G, and C, is folded up into the chromosomes of a cell. How to read it—how to learn the sequence in which the four letters are linked—was discovered by two teams of scientists in the late 1970s. First they divided the chromosomes into parts—like unbinding a book so that its pages are loose. Then they made several copies of one part—one page—at a time. These copies they cut randomly into different lengths. Knowing the length of each fragment and what its last letter was, they could decipher the order of the letters. If the code were written in English, for example, they might have had a four-letter fragment that ends with "e," a two-letter and a five-letter fragment that both end with "r," and a three-letter fragment that ends with "d." Given a one-letter fragment that is an "o," the word is "order."

Today genomes are being sequenced more rapidly than they can be analyzed. Much of the process is automated. Automated, too, is the comparison between one genome and another. Still, each genome hides its secrets. Although we know its code—how the letters of its ATGC alphabet are translated into proteins—we cannot read it like a book. The analogy to a book is, in the end, inadequate, because what is written in the DNA is how to make proteins (and other molecules, like RNA), not how to make an organism.

When the first complete draft of the human genome sequence was published in the journals *Science* and *Nature* in February 2001, *Science* lead author Craig Venter wrote of "a major surprise." There were far fewer genes—no more than 30,000 or 40,000—than the 50,000 to more than 140,000 that had been predicted. That's hardly more genes than the 26,000 in the genome of *Arabidopsis thaliana*, the little weed used as the fruit fly of plant genetics. *Drosophila* itself has 13,000 genes. Venter concluded: "The modest number of human genes means that we must look elsewhere for the mechanisms that generate the complexities inherent in human development."

Nor are there only *genes* in the genome, if by gene we mean the sequence that codes for a protein. Of the string of 3.2 billion ATGCs that make up the human genome, for example, only a tiny fraction—between 1.1 and 1.4 percent—actually codes for the organism's proteins. The coding sequences are scattered throughout a much longer length of DNA in short segments called exons. They are interrupted by noncoding segments, introns, that can be tens of thousands of letters long. The functions of introns are still being discovered, though they are now generally considered to be part of the gene. Introns can increase the transcription of a gene; that is, genes work better with introns.

There are other sequences called promoters and enhancers. These regulate the genes, specifying when a gene is turned on and how active it is. Some of the DNA codes for molecules that guide the editing of gene transcripts—the splicing together of exons and the excision of introns. Other sequences code for regulators or have a structural function. Yet even counting the introns as part of a gene, less than 30 percent of the genome is devoted to genes.

The rest was once labeled "junk" DNA. The name stuck, even after the nature of these sequences was understood. One type of sequence was originally called satellite DNA because it had a rather different composition than the mainstream DNA and therefore formed a satellite band in a standard laboratory analytical procedure. Today such sequences are known to consist of many, many copies of very short sequences and to be important for cell division: they apportion chromosomes when cells divide. But much of what was originally labeled "junk" is made up of transposons, popularly known as jumping genes.

Transposons are genes that can move from one place to another on the chromosome. Although the patterns they form in plant leaves and flowers were well known, it was Barbara McClintock's studies in the colorful maize called "Indian corn" that led to the first reports, back in the late 1940s, that such patterns were caused by genes that move. They remained an abstract concept, like the early twentieth-century notion of the gene itself, until the first one was isolated and cloned decades later. Since then, they have come to explain quite a lot about how genes act—and evolve.

For instance, in 1990 scientists identified the gene that had mutated to give Mendel's peas their wrinkled seed trait. The gene codes for a protein that helps make starch in the seed. The mutation was caused by a transposon: by jumping into the gene's sequence, the transposon made it impossible for the complete protein to be made. Without this protein, less of the sugar in the pea can be converted into starch. With less starch, the pea seed wrinkles as it dries. And in the early 1980s, it was an understanding of transposons—what they are and what they can do—that inspired scientists to harness the soil bacterium *Agrobacterium tumefaciens* to carry new genes into plants.

5

TINKERING WITH EVOLUTION

What distinguishes a butterfly from a lion, a hen from a fly, or a worm from a whale is much less a difference in chemical constituents than in the organization and the distribution of these constituents.

—François Jacob (1977)

By 1804 Julius von Sachs knew that the carbon and nitrogen in a plant were precisely the same as those in a rock: they were not imbued with a vital force just because they resided in a living thing. In 2001 Craig Venter and the other scientists who sequenced the human genome and compared it to the genomes of yeast and fruit flies and mice and maize and the little weed *Arabidopsis* concluded that genes, too, were not "the mechanisms that generate the complexities" of a creature.

Humans have roughly 200 types of cells: blood cells, bone cells, skin cells, liver cells, heart cells, and more. All come from a single cell, the fertilized egg, over a period of years. Yet humans have only 10 times as many genes as the bacterium *E. coli*, itself a single cell. We have twice as many genes as a fruit fly. But a corn plant has a genome as large as ours. Our complement of genes is startlingly similar to that of mice. Set the human sequence and that of the mouse side by side, and only some 300 human genes will have no counterpart in the mouse. Anthropologist Svante Pääbo wrote in 2001 that such cross-species comparisons "make the unity of life more obvious to everyone." They are "a source of humility and a blow to the idea of human uniqueness."

In 1977, well before the first genome was parsed into its A's, T's, G's, and C's, François Jacob had raised that very question in an essay called "Evolution and Tinkering": "What distinguishes a butterfly from a lion, a hen from a fly, or a worm from a whale?" Answering himself, Jacob, a molecular geneticist, made the radical suggestion that evolution was more like tinkering than engineering. Each organism's adaptation to its niche is far from perfect—this fact Darwin himself had been keenly aware of. Yet in the years after *The Origin of Species*, the concept took hold that evolution worked like an engineer designing a machine. Imperceptibly, natural selection was endowed with the ability to optimize. The idea that every bit of a worm or a whale was there for a reason—one understandable through the filter of natural selection—is still cherished by many, if not most, biologists today.

Jacob was saying something quite different. An engineer starts with a plan, and with the materials specified, then designs the product with the appropriate methods to be as perfect as possible. A tinkerer starts with whatever is at hand and makes something that does the job, even if far from perfect. A tinkerer needs no plan—just a supply of spare parts. These, evolution provides in abundance. However life on Earth arose, most likely it did not start twice. Thus each evolutionary change occurred in an organism that already existed. Those organisms, for most of life's history on Earth, were single cells. (Most living things, at least in numbers, still are.)

Today's many creatures show indisputable differences: Hens have no teeth, worms have no feathers, *E. coli* has no flowers, potatoes have no toes. Yet much is the same. Evolution seems to copy a part, then change it a bit, tinkering with it to add a new function.

A cell, whether of *E. coli*, a lion, a fly, or a lettuce, is a jelly-like agglomeration of long molecules, most of which are proteins. The cell is encased in a membrane made of fats, and threaded through with other membranes. At its center is DNA. In higher organisms (from yeast to people) the DNA, in its spindly chromosomes, is inside a separate compartment, the nucleus. In lower organisms (bacteria) the single chromosome is tightly bundled, but not segregated from the rest of the cell.

The cells of higher organisms also house mitochondria. These small structures are descended from bacteria that took up residence inside other cells long ago. Mitochondria have their own DNA, although many of the original bacterial genes have moved into the cell's chromosomes. Inside the mitochrondria are proteins that break down sugar, providing energy for the cell to move and grow. The cells of green plants have an additional kind of cell-within-a-cell, or organelle, the chloroplast. Chloroplasts let plants extract energy from light—a spectacular ability. Chloroplasts are descended from cyanobacteria, a kind of bacteria that learned to harvest light energy more than a billion years ago. They too still have their own DNA. But, just as in mitochondria, many of their original genes have migrated into the nucleus and now reside on the cell's chromosomes.

There are variations on the theme of "cell." Blood cells in mammals fill up with iron-rich hemoglobin, then eliminate their nuclei to focus on toting oxygen from the lungs to every cell. Plants develop stacks of long cells, end to end. These interconnect then die, forming

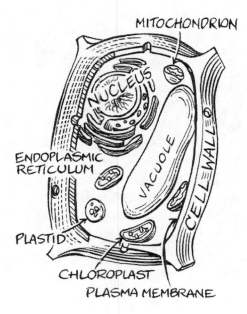

The interior of a plant cell

long straws that pull water out of the ground. But such specializations came late in evolution. From the simplest organism to the most complex, the cells of all living things have much in common. Every time a cell divides, for example, all its DNA must be copied. The copying machine in the bacterium *E. coli* is made up of 10 different proteins: some form the structure of the machine, some latch it onto the DNA, and some link the letters ATGC at a rate of 500 per second. In 20 minutes the copying is done. To copy a human cell's much longer DNA takes about eight hours, with a thousand similar machines.

The process of making a protein is also not much changed from *E. coli* to *Homo sapiens*. Each protein is coded for by a gene, a length of DNA that can be hundreds or thousands (or even hundreds of thousands) of letters long. The gene is first transcribed into RNA (ribonucleic acid), like a television interview transcribed for print. Then the transcript is translated from RNA into protein, as if from Japanese to English. RNA is similar in structure to DNA (deoxyribonucleic acid), except that it contains a different sugar (ribose instead of deoxyribose) and it has only one strand, not two, so it doesn't form the familiar double helix. Like DNA, RNA is made of four building blocks, called bases or nucleotides. Both molecules have the bases A (adenine), G (guanine), and C (cytosine). RNA, however, has U (uracil) in place of the T (thymine) in DNA. The transcript is a long molecule of RNA called messenger RNA. The translation into protein is the job of a number of smaller forms of RNA called transfer RNAs.

All of each gene's DNA—both the protein-coding exons and the mysterious introns—is read and transcribed into an RNA molecule. The chromosome itself is part of this process. Chromatin, the complex of DNA and proteins from which chromosomes are made, has to be remodeled to allow transcription to occur. A large transcription machine must be assembled. This machine unzips the DNA double helix at the appropriate spot and allows the RNA letters room to bind with their partners on one of the DNA strands. As soon as the transcript begins, the leading end is linked to a protein. This protein attracts others, whose job it is to detect the beginning and end of each intron. They cut the transcript there and splice the two exon ends together, kicking out the intron.

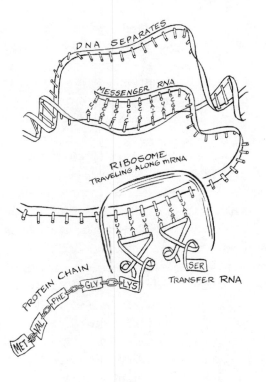

From DNA to protein

When the transcription machine reads a certain signal in the DNA, it stops. A new protein machine then adds a long string of A's to the transcript, familiarly called a "polyA" tail. The tail is the transcript's certificate of accuracy. Throughout the transcription process, layers of proofreaders (proteins again) have been checking for mistakes. The tail is the last rubber stamp. Only when it has a polyA tail can a transcript travel out of the nucleus and be translated into a protein.

In bacteria these two processes—transcription and translation—are completely coupled. Because bacteria are simpler and have fewer (and simpler) genes, they seem to be able to get away without proofreading. In higher cells the transcript, the certified error-free RNA message, goes out of the nucleus and into the cytoplasm—the bulk of the rest of the cell. There the message is read by a ribosome, another machine made of many proteins. The ribosome follows a strict rule of translation: each set of three nucleotides, each codon, in the messenger

RNA is recognized by a matching anticodon at one end of a molecule of transfer RNA. The messenger RNA's UUU, for instance, is recognized by the transfer RNA's AAA.

On its tail end, each transfer RNA has attached to it a particular one of the 20 amino acids, in this case the amino acid phenylalanine. The transfer RNAs take their places along the messenger RNA, lining up the amino acids on their tails in the order specified by the DNA code. The amino acids are linked in a chain and then released from the transfer RNA. The long chain of amino acids then folds up to make a functioning protein. Each kind of protein folds, like yarn into a ball, in its own way, although there are common themes, called "folds," that proteins favor. The sequence of amino acids determines the shape of the protein. The shape of the protein determines what it does.

Curiously, certain protein folds have been especially successful in evolution—all of the hundreds of thousands of proteins fold into no more than a few thousand of these overall structures. And many proteins have proved so useful that their sequences have changed very slowly over the eons. These similarities mean that when a new protein is identified, scientists routinely check its sequence and shape against the databases of those already known. They can often guess what the new protein does from the functions of others that it resembles.

Proteins actually do a surprisingly small number of jobs. They make and break chemical bonds. They bind and transport small molecules, such as hormones, sugars, or oxygen. They bind to nucleic acids (DNA and RNA), to membranes, and to other proteins, giving structure to cells and regulating their chemical activities. Each of these jobs is done by different protein parts, called domains. Protein domains often come in pairs, so a particular protein might bind to DNA with one domain and break a chemical bond in that DNA with another domain. Groups of three or more domains that work together are also common, increasing the variety of jobs a single protein can do. Protein domains are the spare parts of evolutionary tinkering: different proteins have different combinations. By shuffling domains or adding new ones, an old protein can acquire a new function. Yet if you compare a domain from a protein in a bacterium, a plant, and a person, they might look quite similar—both in sequence and in structure.

The similarity among proteins in different organisms reflects not only the underlying fact that the cells of all organisms carry out many of the same chemical reactions, but also the evolutionary relationships among the organisms. Similarity at the protein level doesn't mean that all living creatures are the same—any more than a Model A Ford is the same as an SUV. It means that they're made of parts that do the same kinds of tasks, and it means that organisms evolved gradually. It also doesn't mean that every part in an SUV looks or works the same as its counterpart in the Model A. But because the rules for transcribing genes and translating proteins are the same in all living things, a gene from a bacterium can, with the proper switches added, work in a plant cell. The plant cell will make the very same protein from that gene that the bacterium made.

In the early 1990s Pam Dunsmuir and her colleagues at DNA Plant Technology Corporation, a biotechnology company in California, put a fish gene into a tomato. The experiment has since become a rallying point for protests against genetically modified foods. A Canadian writer, Jennie Addario, reported on one such event at a Loblaws grocery store in Toronto in 2002:

> Suddenly, about 50 protesters march toward the grocery store, chanting, "Hey hey, ho ho, leave our DNA alone." One demonstrator beats a drum. Others hold up homemade placards urging shoppers to "Be GMO Free" and warning them that "You're Eating Genetically Modified Organisms." Some protesters even wear costumes. One young woman sports a hand-drawn cardboard ear of corn that hangs from her neck like an oversized tie. Another is far more creative: she is dressed from top to bottom as a gigantic, plump, red tomato, complete with a large stem and leaf jutting from her head and, bursting out from her vegetable belly, a fish head with eyes, fins, and gills. So it was no surprise that in the next day's edition of *The Toronto Star* a picture of the mutant tomato-fish received prominent play. In the story accompanying the photograph, the reporter, Tanya Talaga, wrote that the protester "said her costume represented the fact that fish genes are spliced into tomatoes to give them a longer shelf life." The impression given to the reader by this statement and the rest of Talaga's story was that all the tomatoes in grocery stores, including those at Loblaws, were swimming in fish genes. But that is simply not the case. In fact, there are no supermarket tomatoes anywhere that are packed with fish genes.

The "mutant tomato-fish" is nothing more than an urban legend, writes scientist Alan McHughen in his book, *Pandora's Picnic Basket: The Potential and Hazards of Genetically Modified Foods*. It is a story made of "shadows" to "incite public fear by creating an unnecessary scare."

Dunsmuir's experiment did not show the effect she was looking for, and a tomato containing a fish gene was never grown or commercialized. Nor did increasing shelf life have anything to do with it—that was another tomato, Calgene's Flavr Savr, which was the first genetically engineered whole food to come to market. First sold in 1993, the Flavr Savr contained no fish genes. In fact, it contained no foreign genes at all. The modified gene was a normal tomato gene that coded for an enzyme, polygalacturonase (PG), involved in the fruit ripening process. The researchers had copied the PG gene, flipped it, and reinserted it in reverse order. "The resulting 'antisense' PG gene, by some unknown mechanism," wrote one of the researchers, Belinda Martineau, in her book *First Fruit: The Creation of the Flavr Savr Tomato and the Birth of Genetically Engineered Food*, "shut down native PG protein production in the engineered plants." Ten years later we know that the "shutdown" is a normal plant response that protects from invasion by viruses.

According to Martineau, the Flavr Savr "sold like hotcakes." "Bert Gee, the owner of State Market, resorted to limiting customers to two Flavr Savr tomatoes a day," Martineau writes. He sold "gift packs— 'four tomatoes in a box to give to friends and family.'" A waiting list formed of grocers wanting to stock Calgene's product. But Calgene couldn't deliver. Belatedly they discovered that they "had to handle vine-ripened Flavr Savr tomato varieties, it seemed, just as gently as any other, conventionally developed vine-ripened fruit." The assumption that a tomato slow to rot would stay firm enough to pack and ship "wasn't panning out." Their $2-per-pound tomatoes cost the company $10-per-pound to deliver. Flavr Savr was off the market by 1996. Calgene, near bankruptcy, was sold to Monsanto.

The tomato with the fish gene, on the other hand, never made it out of the lab. Although the gene itself worked properly—it produced the same protein in the tomato as it did in the fish—the protein did not have the desired effect, which was to protect the tomato plant from frost in the field and the tomatoes from cold damage on the way to market or at home in the refrigerator. The gene came from a type of Arctic flounder; the protein it encodes keeps the fish from freezing in icy waters. When cells freeze, the transition from water to ice breaks them apart just as surely as leaving a bottle of beer in the freezer breaks the bottle. Yet many organisms, both animals and plants, avoid such severe freeze damage. Seeds, all trees, and many bushes do, as well as salamanders, snakes, insects, and some fish. Just as we use salt to de-ice airplanes or roads, they use high concentrations of salt or other substances that keep water from freezing until the temperature is colder than normal.

However, 30 years ago scientists working in the Antarctic found fish that fought freezing a different way. Special proteins in their blood allow them to survive temperatures that would burst even salt-rich cells. These antifreeze proteins lower the freezing point through a mechanism that, unlike the salt method, does not depend on their concentration. As the ice crystals are forming, the proteins bind to the prism faces of the ice. The ice crystals grow more slowly and so do not disrupt the fish's cells.

Such antifreeze proteins have since been found in flat fishes, sculpins, sea ravens, smelt, herring, eel pouts, northern cods, and wolf-fishes. They are found in mealworms and spruce budworms, and in the larvae of the fire-colored beetle. Winter rye has them, as do carrots and bittersweet nightshade, a plant in the same genus as tomato, *Solanum*. All of these proteins work in the same basic way, although how they are related to one another is not clear. One of the fish proteins appears to have evolved from a protease, an enzyme that breaks down proteins. Others, isolated from winter rye, originated from chitinase genes recently enough that some of them still retain their earlier function, which is to break down polysaccharides, a kind of sugar polymer.

The antifreeze proteins in winter rye and carrots and nightshade are not as effective as those in fish or insects, and so scientists have continued their efforts to insert new antifreeze genes into plants. The goal of most such experiments is to protect crops better from untimely frost. A gene from the fire-colored beetle, for instance, was recently engineered into *Arabidopsis thaliana*, the laboratory plant now popular with plant geneticists. The insect protein worked quite well, lowering the temperature at which the plants froze.

Why did it work? First, of course, the rules for transcribing and translating DNA into proteins are everywhere the same. The DNA codon AAA always stands for the amino acid phenylalanine; CTT always means glutamine. The amino acid sequence is the same whether the gene resides in a fish or a beetle or *Arabidopsis* or a tomato. Among organisms like these, whose evolutionary paths diverged so long ago, there are of course subtle differences. Each has its own way of saying when and where a gene will be expressed. They differ in how often they use a certain codon—with 64 possible triplets but only 20 amino acids, two or more codons can be synonyms. But the triplet code itself is universal. A gene directs the construction of precisely the same protein in whatever organism it finds itself in. The second reason the new protein works is that a protein's task, always and everywhere, is determined by its sequence and how it folds.

The fact that many different kinds of proteins evolved to prevent freezing in many different organisms illustrates just the kind of tinkering Jacob had in mind. Some of these proteins are efficient at preventing freezing, some less so. But they are all simply proteins. It is as impossible to attach the quality of being an insect to a protein as it is to attach the quality of being a car to a spark plug. There is no way a gene from a fish could make a tomato "fishy." When the fire-colored beetle gene is expressed in *Arabidopsis*, the plant is still *Arabidopsis*. Only now it can express one extra protein, one that changes how ice crystals form.

Whether the antifreeze protein will do for *Arabidopsis* (or a tomato) what it did for the beetle (or the fish) from which the gene was taken is more difficult to predict. It has to be tested, as Dunsmuir was doing. Each organism modifies its proteins a little differently. These modifications can determine the fate of the protein within the cell, as well as what partners the protein works with. As organisms become more complex, their proteins become more and more exclusive. They are pickier about where they carry out their functions and whom they partner with. They will work only in the nucleus or only in the mitochondrion or only on the cell's outer surface. Or the gene that encodes a protein will be expressed in one cell type, but not another.

For example, hemoglobin binds oxygen in both animals and plants. In animals, hemoglobin proteins are made in red blood cells that circulate in the blood stream to bring oxygen to cells far from the lungs. In legumes, like peas and beans, they are used for a different purpose: to create the perfect environment for the nitrogen-fixing bacteria in the plant's roots. The enzyme that converts nitrogen from the air into compounds the plant can use to grow (so that it does not need nitrogen fertilizer) can't function in the presence of oxygen. So the plant builds an oxygen-free home for this beneficial bacterium, lining special root nodules in which the bacteria live with hemoglobin molecules to keep oxygen away. The plant and animal hemoglobin molecules are not identical—they've drifted apart over evolutionary time. But their ability to bind oxygen is much the same and their similarity in structure is still recognizable.

So to paraphrase François Jacob, the difference between a person and a pea is less a difference in their genes than in how, where, and when those genes function and the proteins are deployed. The difference between a fish and a tomato is not so much in what their proteins do, but in how much of each is made, where it is made, and how it interacts with its partners.

With this new understanding of proteins comes a deeper understanding of evolution. Evolution once seemed quite linear—and human beings were obviously as near to perfection as any organism had come. But some complex animals, such as horseshoe crabs and coelacanths, have persisted unchanged for eons. We call them, quaintly, living fossils—as if they ought to be extinct. Indeed, many species of complex animals are extinct, even as simpler organisms have persisted. Modern bacteria are not very different from fossil bacteria. They have the same basic single-celled layout, despite many specializations and diversifications. Newer life forms do not replace older ones because they are more complex, higher, or better adapted. Instead, the evolution of life on Earth is marked both by increasing complexity and by boisterous diversity. The diversity is easier to understand than the complexity. Many organisms can be derived from an original type with just a little bit of tinkering. From ordinary fish, living deep in caves where light never penetrates, come blind fish without eyes.

But where does the complexity come from? If evolution begins with an existing fish and then adapts it, how can something essential ever change? How can a gene be rearranged to give rise to a new protein with a new function or a new pattern of expression if the original protein is one the organism can't live without?

One answer is to have spare copies of genes. Often a single gene is copied—then copied again and again, with all of the duplicates staying in place, side by side. Sometimes a whole section of chromosome is doubled—or even the whole genome. To copy a whole genome is in fact quite simple. The machinery is already in place, because every time a cell divides, all of its DNA—the whole genome—is copied. Sometimes both copies go into one cell by mistake when the cell splits into two daughter cells. Occasionally, the daughter cell with a doubled genome is a germ cell—a sex cell—and so the doubled genes are handed down to the next generation. This, incidentally, is how colchicine works, the chemical from the autumn crocus with which plant breeders created triticale and the seedless watermelon. When the genome is copied, colchicine keeps both sets together.

Extra copies of genes, chromosomes, or whole genomes are made in the simplest organisms and in the most complex. But the copies tend to persist in eukaryotes, organisms whose cells have nuclei, much more often than they do in prokaryotes, the bacteria. Bacteria have a handful of duplicated genes; humans, mice, and corn plants (all eukaryotes) have many thousands.

Just why extra copies are made and why they persist, swelling genomes, is not well understood. But there is no doubt that genomes grow in the course of evolution by duplicating their DNA. Once a gene is doubled, changing the protein is easy: one copy of the gene supplies the original protein; the other copy can change. A duplicated gene can be lost—and they often are—or it can make itself essential by acquiring a new function—and they occasionally do. Many, if not most, genes in plants and animals belong to clusters, or families, of related genes. Some gene families are small, with as few as two members; others are huge, with hundreds or even thousands. Each family arose by copying and tinkering with a previous DNA sequence.

If copying is the key to complexity it should be no surprise that the genomes of animals and plants are full of transposons, or jumping genes. But when Barbara McClintock announced in 1948 that genes could move from one place to another, it was tantamount to saying the kitchen could occasionally move into the attic. Geneticists—McClintock among them—had been mapping genes for decades. They knew very well that genes maintained their locations—maps couldn't be made if genes moved around all the time. Genes had fixed locations, like beads on a string, and their order along the chromosomes could be counted on generation after generation. Occasionally chromosomes exchanged parts—this was called a translocation—but these were unusual events. Genes didn't jump. Yet McClintock reported that they did.

McClintock began her work at Cornell University in the early 1920s, at the same time as George Beadle was beginning his studies there of corn (maize) and teosinte. McClintock was curious about the relationship between chromosomes and the inheritance of traits. To understand it, she too worked with corn.

From the account of one of McClintock's biographers, you would think that the genetics of corn was a gentlewomanly pursuit: she "watches the plants grow over the summer, and spends the long quiet winters analyzing the results." A scientist who worked with McClintock reported a very different experience: "Doing maize genetics is hot and sweaty work—sunup to sundown—sometimes in temperatures over 100 degrees. And then there's keeping track of what crosses have been done and what crosses must be done the next day. Every plant and every cross must be recorded before the next day's crosses are planned." She remembers being so exhausted by the end of two months of 5:00 A.M. to 2:00 A.M. days that she seriously considered abandoning the project. Even McClintock admitted that it took her months to recover from each summer's work.

When corn flowers, pollen bursts from the tassels at the plants' tops. Millions of grains swirl through the air, to be snagged by the long silks protruding from the female flowers, the ears. Controlling this boisterous mating process takes extreme care. Each plant is given a number. Each ear is covered by a bag. When an ear is ready to be fertilized, the bag is briefly lifted and pollen from the desired male is brushed onto the silks of the ear. The tracking numbers of the plants used as the male and the female in the cross are written on the bag. (In today's genetics labs, these numbers are in the form of barcodes.) Hundreds and hundreds of crosses are done every day at the height of the season. Thousands are done in all. A month or so later, each mature ear is tagged and collected. Then the serious work begins.

A way to visualize the chromosomes of maize had been McClintock's first major contribution to science. Corn's chromosomes had been counted—there were 10—but in the 1920s no one could consistently tell one from the next. While a graduate student, McClintock found that the chromosomes in male reproductive cells were the best ones to examine, because they were long and rather stretched out. She adapted a staining technique that had just been developed using a red dye called acetocarmine. The stain made the structure of each chromosome clear. McClintock gave each one a number, distinguishing them by size, by where they attached to each other, and by the pattern of various knobs each one bore. She received her Ph.D. in 1927 and stayed on at Cornell. There she and a student, Harriet Creighton, soon succeeded in connecting the behavior of genes with the behavior of chromosomes. Their paper, published in 1931, has been called one of the "classic papers in genetics," a "landmark in experimental genetics," even a "cornerstone." Its results "are so strong and obvious that their validity cannot be denied."

This paper made McClintock famous. In 1936 Lewis Stadler, who was then carrying out some of the first mutation breeding experiments on barley, persuaded the University of Missouri to offer her an assistant professorship. There she began to study the behavior of maize chromosomes that had been broken by X-rays. When George Beadle, in the 1940s, was exposing the orange breadmold *Neurospora crassa* to X-rays—work that would ultimately lead to his Nobel Prize-winning "one-gene, one-enzyme" hypothesis—he asked McClintock to identify the mold's chromosomes. Much later, in the 1950s, the National Academy of Sciences became concerned that the introduction of high-yielding hybrid corn to Central and South America would result in the disappearance of local varieties, or landraces, a concern that has resurfaced today in a different guise. The Academy asked McClintock to visit field stations to teach local biologists how to classify types of maize by looking at their chromosomes. She was considered the world's foremost expert on the structure of the chromosomes of corn.

But the work that most deeply engaged her for more than 50 years, and for which she finally received the Nobel Prize in 1983, at first drove her into obscurity. It began with broken chromosomes. McClintock devised a method for bringing broken chromosomes together from each parent. She saw that the broken ends stuck together, forming a chromosome with two centers—a dicentric chromosome. Normally when a cell divides, it finds a chromosome's centromere, a special structure at its center, and pulls the chromosome apart at that point. A dicentric chromosome has two centromeres. When the cell divides, it is caught in a tug-of-war. It breaks as the cell attempts to distribute the two centromeres to the two daughter cells. Because it breaks at random, one daughter cell sometimes receives two copies of a chromosome segment (and the genes it carries) and sometimes none.

McClintock saw that the offspring of such plants suddenly showed many new mutations. A startling number of them were the kind called unstable mutations, in which the mutant gene often reverts to its normal function while the plant is growing. These kinds of mutations cause the stripes or spots in decorative Indian corn. Such variegation is also familiar—and highly prized—in flowers like snapdragons and roses.

McClintock found that one mutation seemed to cause a chromosome to break, not at random, as she was accustomed to, but at the same place each time. She began to study this curious gene, which she called *Dissociation*. Her next discovery was truly momentous: *Dissociation* could move. In the margins of an unpublished paper on her first *Dissociation* work are these comments in McClintock's hand: "At the time, I did not know that *Dissociation* could change its location. Realization of this did not enter my consciousness until late this spring, following the harvest of the greenhouse crop." The spring was that of 1948 and the greenhouse crop was grown in the winter of 1947.

Long before McClintock began her work, Rollins Emerson, George Beadle's advisor, had studied the unstable mutations that cause the col-

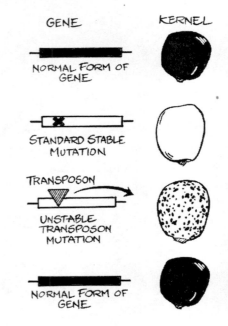

The effect of a transposon, or jumping gene, on the colors of a corn kernel

orful patterns in Indian corn. Previous geneticists had simply thrown up their hands when confronted with this peculiar behavior; some dismissed them as "sick" genes from which nothing could be learned. But using a corn variety whose kernels were striped dark red, Emerson found that the genes' behavior wasn't so bizarre after all. It could be explained quite well if some kind of inhibitor had gotten stuck to the gene, keeping it temporarily from working. If the gene were needed for colored kernels to be made, the inhibitor would cause colorless stripes. But if the inhibitor fell off now and then while the plant was growing, the gene would work normally: streaks or spots of color would appear. Emerson couldn't imagine what this inhibitor might be.

Some years later Marcus Rhoades, a lifelong friend of McClintock's, learned something else about unstable mutations. When he crossed a plant carrying a mutation in a pigment gene (a standard, stable mutation studied by maize geneticists for years) to a variety called Black Mexican Sweet corn, the mutation suddenly became unstable. The instability was caused by a gene he called *Dotted*. Rhoades's discovery meant that a mutation's stability could depend on the presence or absence of a completely different gene.

McClintock was familiar with all of this work when a new mutation cropped up in her plants in which *Dissociation* was breaking chromosomes. McClintock noticed an odd-looking kernel. It was largely colorless, but had little colored spots, much like the dots that Rhoades had studied. McClintock knew that *Dissociation* could move from one place to another in the chromosome. She guessed now that it might occasionally jump into, and disrupt, another gene.

By the time McClintock had tested—and verified—this prediction, she knew the answer to Emerson's question. The inhibitor was a transposable element (in this case, *Dissociation*) that could jump into and disrupt the function of a gene. If its activator gene was present (in McClintock's plants it was called *Activator*, while in Rhoades's plants it was *Dotted*), then the transposable element—today called a transposon—could jump out again.

McClintock's first major public lecture on transposition was at a Cold Spring Harbor Symposium in 1951. She told an auditorium full of her fellow geneticists that not all genes were fixed in their positions on the chromosomes. Some, like *Dissociation* and *Activator*, could jump to new places on the same chromosomes. They could even transpose to a different chromosome. Most of her colleagues were simply puzzled.

By then, McClintock enjoyed an extraordinary reputation as a geneticist. She'd been one of the first women elected to the prestigious National Academy of Sciences. The eminent geneticist Sturtevant, maker of the first gene map, when asked what he thought of McClintock's work, reportedly replied, "I didn't understand one word she said, but if she says it is so, it must be so!" Lewis Stadler, who had recruited McClintock to work at the University of Missouri in the 1930s, pronounced it "the most amazing thing." He added that because of her reputation and great observational skill he had "no choice but to believe her." Others were less generous, but no less confused.

Because McClintock was on the scientific staff of the Carnegie Institution of Washington's Department of Genetics, she didn't have to care what her colleagues thought—or whether the government would fund her research. Her salary was paid. She worked by herself, growing and analyzing corn. She simply did what fascinated her.

Transposons were discovered in the early 1960s in bacteria. In the 1970s and 1980s they were discovered in fruit flies, yeast, worms, fungi, mice—and finally humans. Few organisms lack them. Today we know that genomes are stuffed with transposons and their relatives, called retrotransposons because they multiply outside of chromosomes and then reinsert themselves. Almost half of the human genome and three-quarters of the maize genome consist of nothing but transposons and retrotransposons.

Why did it take so long to discover transposons? Many writers have looked for the answer to this riddle in McClintock's reclusive character and her supposed inability to communicate. Her first long paper on transposons is compared uncharitably to Watson and Crick's 800-word note—which one critic described as "tight as a sonnet"—on the structure of DNA. Watson and Crick's contribution was hailed almost immediately: they were awarded a Nobel Prize a mere nine years later. Three decades would elapse before McClintock received her Nobel Prize.

But the answer doesn't lie in the person. It is both straightforward and mysterious. The much slower acceptance of McClintock's discovery, compared with that of Watson and Crick, has a parallel in Mendel's obscurity and Darwin's popularity a century earlier. Darwin's theory of evolution was immediately recognized as important, while Mendel's laws of heredity weren't widely integrated into evolutionary thinking for 50 years or more. Darwin's theory explained things that were easily observed: birds resemble each other in general ways, but each bird is a bit different. It wasn't at all clear initially what Mendel's rules of inheritance had to do with the differences that lead to new species.

Equally unimaginable a century later was the importance of jumping genes. Mendel's laws were by then the central paradigm of genetics. The mapping of genetic loci through the study of mutants was proceeding apace. There was plenty of evidence that genes had fixed locations. Geneticist Ledyard Stebbins' book *The Basis of Individual Variation,* published in 1950, acknowledges that there are duplications, inversions, translocations, and deletions in chromosomes. Still, the book reflects the prevailing view that these "are not the materials that selection uses to fashion the diverse kinds of organisms which are the products of evolution." Instead, he concludes, the majority of evolutionary changes are due to classical genetic point mutations: an A substituted for a G, a T, or a C.

Now another half century has elapsed. In every genome scientists have explored, they find themselves knee-deep in transposons, most of them defective in one way or another—the bones of dead transposons. Those that are not defective are commonly silent, kept inactive by reversible mechanisms. The result is that neither genes nor transposons normally move; chromosome structure is maintained and the transposons are undetectable.

Still, it seems that having an abundance of transposons goes along with being quick at the game of evolution. When the genomes of closely related animals (such as mice and humans) or plants (such as sorghum and maize) are compared, the difference between them lies in the linear relationships among the genes. Often the order is the same for long stretches. Elsewhere, a gene is missing from one genome while persisting in the other. Or a small group of genes occurs in the opposite order in one genome, as if a chromosome segment had been flipped around. Or a gene—or group of genes—has moved from one chromosome to another. Such rearrangements frequently start and end with a transposon. It is hard to escape the conclusion that transposons are central to rearranging chromosomes and moving genes.

Genes can change, they can duplicate and delete, and genomes scramble. It is increasingly evident that what genes do depends more on *what* they are than *where* they are—although both a gene's immediate neighbors and its general genomic neighborhood can influence its expression. But evolution takes a long time—like the movement of tectonic plates. The evolution of a plant is measured in millions of years, not in the months it takes to grow a crop of corn. McClintock saw transposition on this very short time scale only because her experiments with broken chromosomes had awakened sleeping transposons.

It is usually to an organism's advantage to keep its transposons carefully out of action—silent, asleep. Jumping genes might well power evolution, but they can also do a great deal of damage. When transposons move, they almost always increase in number. The jumps occur just after the chromosome divides, in preparation for cell division. One copy of the transposon stays in its original place, while the other copy moves to a new location—sometimes on the same chromosome, sometimes on a different one. Each time a transposon moves, it creates a sequence like itself somewhere else in the same genome. When the transposon alights in part of a chromosome that has not yet been replicated, it doubles again, along with the rest of its new chromosome. Then there are three copies of the transposon: one at the original location and two at the new spot.

During the crossing-over phase of cell division, chromosomes exchange parts, or recombine, when the same sequences on two chromosomes line up perfectly with each other. In this way, a new combination of grandmother's and grandfather's genes is passed on to the next generation. Normally the chromosomes exchange equal segments, because the same gene is located in the same place on every chromosome. Transposons can disrupt that pattern. Copies at different locations can—and very occasionally do—recombine with each other, causing a translocation, an exchange of parts between different chromosomes. The genes get shuffled. Or a whole segment of the chromosome, bounded by transposons, is duplicated or simply moved to another chromosome.

Transposons often delete portions of themselves as they try to move, damaging and immobilizing themselves. That's why genomes are full of transposon bits and pieces, often just the very ends. Transposons undergo bursts of activity, then burn out by accumulating defective copies. The *Dissociation* transposon McClintock first identified was such a defective element, made up of two transposon ends without a middle. *Activator*, the gene that made *Dissociation* jump, turned out to be a complete, functional transposon. It supplies *Dissociation* with the enzyme, called a transposase, that it needs to cut itself out of the DNA and move to a new location.

A transposon that is silent or asleep doesn't transpose—but neither does it damage itself, so it can survive to become active another day. There are many ways to silence a transposon. A common one in higher plants and animals is one that doesn't affect the transposon's genetic structure, so it is called epigenetic—on top of the gene. Small chemical groups, called methyl groups, are added to the DNA. Methyl groups can be put on and taken off without damaging the DNA. But methylation profoundly affects the way that proteins bind to a gene. The transposase enzymes, for example, that move McClintock's transposons simply don't bind to DNA that is covered by methyl groups. So as long as the transposon is methylated it won't jump.

By bringing together broken chromosomes and initiating cycles of chromosome breakage, McClintock triggered a process that takes off the methyl groups and reactivates the transposons. Years after her initial discoveries, McClintock showed that this chromosome breakage process could wake up Rhoades's *Dotted* element too, one she'd never seen active in her plants.

Today we know that the same process can be triggered by irradiating plants with X-rays or gamma rays, or simply through ordinary tissue culture, those methods of regenerating whole plants from clumps of callus cells or from protoplasts. The kinds of crosses that Luther Burbank did between very distantly related plants often have the same results. All of these procedures—wide crosses, tissue culturing, irradiation—have been used extensively in conventional plant breeding. By trial and error, breeders found ways to speed up plant evolution by disturbing and rearranging genomes.

6 MAKING A CHIMERA

*In the history of biotechnology, therefore, lies the
story of the twentieth century wrestling with the concept of life.*

—Robert Bud (1993)

According to botanist John Torrey, the first artificial or unnatural
crops were Julius von Sachs's corn and beans, growing in their stop-
pered jars of fortified water on his sunny windowsill in the 1860s.

Others date the birth of unnatural crops to the 1970s, when Canadian plant breeders created canola oil. Keith Downey and Baldur
Stefansson were working on the brassica called rape, thought to have
originated from an ancient cross between a cabbage and a turnip. The
oil pressed from rape seeds was known to be the finest lubricant for
steam engines in ships, but when the price of diesel plummeted after
World War II rapeseed oil went out of fashion. It had also been used
for cooking in Canada and Europe until health researchers linked the
erucic acid in the oil with heart disease. The Canadian government
suggested that growing rape be banned.

Yet rapeseed was a useful crop on the prairies, where little else was
harvested but wheat. Downey and Stefansson embarked on a project
to breed a variety of rape whose oil would contain less of the harmful
erucic acid. They made thousands of crosses between different varieties. To identify variants with a different oil composition, Downey la-
boriously sectioned each seed with a scalpel, then analyzed the oil
content of one part by gas chromatography, a technique borrowed

from the petroleum industry. If the level of erucic acid was low, he and Stefansson would plant and grow the remainder of the seed. By 1974 they had a new rapeseed oil in which the erucic acid was replaced by oleic acid. The new oil—soon dubbed Canola, for Canadian Oil—had the lowest content of saturated fats of any oil on the market and quickly became a popular health food.

Brewster Kneen, author of *The Rape of Canola*, believed, according to a newspaper interview, that the "violent intervention of Keith Downey's scalpel—the 'technology' he introduced—was, in fact, symbolically and practically the beginning of commercial genetic engineering; the deliberate reconstruction of living organisms to create novel life forms for purely human (and commercial) purposes."

But despite Kneen's dramatic account, what Downey and Stefansson did to create canola was hardly new or unnatural. They used the scalpel to separate the nutritive endosperm, where the oil is stored, from the embryo, which grows into the next-generation plant. Both the endosperm and the embryo have the same genes; a mutation that affects the oil composition of the endosperm will almost certainly be carried by the embryo and show up in the resulting plant. Downey and Stefansson's innovation was to analyze the oils with a gas chromatograph—now standard equipment for assessing food quality—in their hunt for spontaneous mutations in the genes that determine the composition of the oil in the seeds. When they found an endosperm with the kind of oil composition they wanted, they planted the accompanying embryo.

Apples could be said to be far more unnatural than that. They are often described by their growers, even, to be "the products of 'artificial trees,'" notes writer Sue Hubbell in *Shrinking the Cat: Genetic Engineering Before We Knew About Genes*. All cultivated apples come from branches of one tree grafted onto the trunk and roots of another. Many new apple varieties first arose as branch sports, just a branch or two bearing apples of a better color, taste, or size. Growers propagate such mutants by cutting the branches into slips and grafting them onto new rootstocks. You can't just plant the apples' seeds. The seed of a prize apple will grow into a tree whose fruit might be nothing like its parent's. Sexual reproduction—the source of the seed—mixes up the genes, often obliterating the desirable color, taste, or size. Grafting, on the other hand, is a type of cloning: the fruit-bearing part of the new tree will be genetically identical to the original branch, and its fruit will have all the desired qualities. And while apple rootstocks are generally apple varieties themselves, the rootstock for a grafted pear tree can be quince, a different genus and species altogether.

Such creations that mix two or more species in a single tree are called botanical chimeras, from the mythical beast that mixed bits of lion, serpent, and goat. These chimeras have been made for at least 3,000 years, the age archaeologists assign to the art of grafting.

Stanley Cohen and Herbert Boyer were familiar with the botanical use of the term chimera when they began the collaboration that would result in their chimeric DNA molecules, ones in which DNA from two different species were spliced together. The technique they developed— at about the same time as Downey and Stefansson developed canola— is the essence of recombinant DNA technology. It makes the modern genetic modification of food plants possible.

Cohen at Stanford University was interested in plasmids, the tiny circular extra chromosomes in bacteria. He had been studying plasmids that carry drug-resistance genes since he arrived at Stanford in 1968; by then 60 to 80 percent of bacteria were showing resistance to more than one antibiotic drug. Cohen hoped to learn how bacteria used plasmids to pass around their resistance genes so that the process could be slowed or stopped. "One of the goals of my research," he said in a 1992 lecture, "was to map and characterize plasmid genes. To do this I wanted to be able to take plasmids apart and put them back together again one segment at a time."

In 1971, Leslie Shiyu, a first year Stanford medical student working in Cohen's lab, discovered that when bacteria were treated with the chemical calcium chloride, they took up plasmids, replicated them, and passed them on to their progeny. Cohen reported these results at a conference in Hawaii. "Later that day I listened with excitement," he recalls, as Herbert Boyer, then at the University of California at San Francisco (UCSF), described experiments in his lab using enzymes to cut apart and glue back together segments of DNA. "That evening," Cohen recalls, "Herb and I, joined by several other friends and colleagues, had a late night snack at a delicatessen across from Waikiki Beach." Over hot pastrami and corned beef sandwiches Cohen proposed a collaboration.

The experiment involved two plasmids from Cohen's lab: one plasmid conferred resistance to the antibiotic tetracycline; another, larger plasmid carried resistance to the antibiotic kanamycin. The experiment also required two enzymes, isolated in Boyer's lab. One, called *Eco*RI, was a restriction endonuclease; like molecular scissors, it could cut the plasmid DNA apart. The various cut fragments could be glued back together with the second enzyme, called a ligase. The tetracycline-resistant plasmid would be cut open, offered the piece of the other plasmid that contained the kanamycin-resistance gene, and glued shut. The recombined, or recombinant, plasmid would then be introduced into bacteria, and they would be exposed to the two antibiotics. If the bacteria lived, it meant that the tetracycline-resistant plasmid had picked up and copied, or cloned, the fragment of DNA containing the kanamycin-resistance gene.

Cohen recalls: "The months in early 1973 were a period of almost unbelievable excitement. The strategy worked even better than we could have expected and on most days there was a new result and a new high. Plasmids were isolated at Stanford, transported to UCSF where they were cut by *Eco*RI and analyzed, and then transferred back to Stanford where the DNA fragments were joined and the plasmids were reintroduced into bacteria." Annie Chang, a research technician in Cohen's laboratory, isolated the plasmid DNA and introduced it into the bacteria. Soon the team was joined by John Morrow of Stanford and Howard Goodman in UCSF in an effort to clone animal (in this case, toad) genes instead of bacterial genes in a plasmid. That strategy worked too, as did one that Cohen and Chang, working with Stanford's Robert Schimke, devised a few years later: they inserted a mouse gene into *E. coli* and showed that the bacterium could express the protein the mouse gene coded for.

They called their recombinant DNA molecules chimeras, Cohen wrote, "because they were conceptually similar to the mythological Chimera . . . and were the molecular counterparts of hybrid plant chimeras produced by agricultural grafting." By 1975 the gene splicing technique was being called plasmid engineering or molecular cloning or genetic engineering or—Cohen's choice—genetic manipulation. "Genetic manipulation," he wrote in *Scientific American,* "opens the prospect of constructing bacterial cells which can be grown easily and inexpensively, that will synthesize a variety of biologically produced substances such as antibiotics and hormones, or enzymes that can convert sunlight directly into food substances or usable energy. Perhaps it even provides an experimental basis for introducing new genetic information into plant or animal cells."

Boyer soon used the technique to produce human insulin in factories made up of colonies of transformed *E. coli.* He founded the company Genentech in 1976, and by 1982 genetically engineered insulin was on the market. It cost much less and was purer than the previous form of the drug. Most important, it was *human* insulin. The insulin formerly sold had been extracted from pigs; it could cause allergic reactions because it was a protein foreign to the human body. Soon other drugs were being produced in genetically engineered bacteria or yeast, such as human growth hormone, interferon, and tissue plasminogen activator, a drug given to stroke victims.

Paul Lurquin writes in his book *High Tech Harvest* that Cohen and Boyer's experiments "marked the birth of biotechnology." But the term "biotechnology" had in fact been in use since 1917. Historian Robert Bud traces it from the concept of zymotechnology, the study of yeasts, molds, bacteria, and other fermentation agents for making bread, wine, cheese, and beer. The term "zymotechnology," from the Greek for "leaven," was coined in 1697. The word "enzyme," which shares the same root, was coined in 1897 by German chemist Eduard Buchner to describe those fermentation agents. The first industrial superstar of zymotechnology's twentieth-century daughter, *bio*technology, was coincidentally an antibiotic: penicillin, which was worth $268 million in 1955. Other stars include the rest of the antibiotics, vitamins B12, B2, and C, corticosteroid hormones, and the enzymes that make high-fructose corn syrup, which by 1974 was sweetening almost every soft drink.

The connection between zymotech and biotech is clear in the naming of the first food ingredient to be produced by gene splicing, chymosin, a protein necessary for cheesemaking. Until 1990 cheesemakers used rennet, made from the membrane of the fourth stomach of a calf; chymosin is the active ingredient. Genetically engineered chymosin was called fermentation-produced chymosin or FPC; it was declared kosher, halal, and vegetarian in 1994. Five years later it had an 80 to 90 percent market share. As Ralph W. F. Hardy of the National Agricultural Biotechnology Council testified before Congress in 1999, "Any person who eats cheese in Canada and the U.S. has been eating a food whose processing involves a transgenic food product." Transgenic is yet another way of saying genetically modified. A transgene is a gene that has been moved from one organism into another using molecular techniques.

The Cohen-Boyer experiments were a milestone because they showed that DNA could be recombined in a test tube. They were not, however, the first time a bacterial cell had been genetically engineered.

In the early 1950s Norton Zinder, then a graduate student in the laboratory of Joshua Lederberg at the University of Wisconsin, showed that viruses could move genes from one bacterial genome into another. As a graduate student himself in the late 1940s Lederberg had discovered sex in bacteria. Until then it was thought that rather than mating and mingling the genes of father and mother a bacterium simply doubled its own genes and split into two genetically identical daughter cells: they were clones. In animals, only identical twins are clones. But plants commonly clone themselves in nature by such devices as underground runners. Stands of poplar, for example, are all clones. And cloning is commonly used to propagate food crops. Potatoes grown from potato eyes are clones, as are raspberry shoots raised from suckers, or apples from grafted branches. Indeed, the word "clone" comes from the art of grafting: it is Greek for "twig."

Lederberg found that bacteria were not all clones. Bacteria have a single circular chromosome consisting of several thousand genes. But most bacterial cells also contain those tiny circles of DNA called plasmids, a name Lederberg coined. Although they replicate at the same time as the bacterium's own chromosome, these plasmids generally remain separate, pursuing their own independent lifestyle. Some plasmids very occasionally integrate themselves into the bacterial chromosome, melding the two into one larger circle. And, as Lederberg discovered, plasmids often carry genes that tell the bacterium to create a connection—a tube or bridge—to another bacterium, through which the plasmid transfers itself. If the plasmid is integrated into the bacterial chromosome when it comes time to transfer, then it drags part or all of the bacterial genes along with it.

Once inside, the new bit of DNA finds the corresponding stretch of DNA in the resident chromosome and replaces it. This process is called homologous recombination, from the Greek for "agreeing in proportion." It is the same process by which the traits of parents, plant or animal, are mixed and redistributed to their children in sexually reproducing organisms: DNA molecules with nearly identical sequences exchange parts. It occurs in nearly all living cells. It is the basic purpose of sex.

Lederberg found several different strains of *E. coli* that could have sex this way (although some within the same species could not). He also found he could mate *E. coli* with several species in the genus *Shigella* and with one species of *Salmonella*, creating hybrid bacteria— and raising the question, once again, of how to define a species. Lederberg published his discovery of bacterial sex as his Ph.D. dissertation. In 1958 it brought him a Nobel Priz; years later this work was also hailed as the birth of biotechnology.

Zinder made an equally astonishing observation: he could transfer a genetic trait from one bacterium to another even when the two did not touch. The genetic engineer in this case is a kind of virus called a bacteriophage, or "eater of bacteria." Before recombinant DNA techniques were invented, bacteriophage—or simply phage—were a favorite for genetics experiments because they are so simple. They have as few as 4 genes, though more commonly they have 50 to 100. They are easy to grow in large numbers because they clone themselves inside the bacterial cell. One infecting phage makes hundreds or thousands of copies of itself inside a single bacterial cell—and a teaspoon of nutrient solution holds 100 million or more bacteria.

Zinder found that a phage could pick up hitchhiker genes from one bacterium and transport them to another. The bacterium he was studying was *Salmonella typhimurium*, a germ that gives mice the equivalent of typhoid fever. He and Lederberg had two mutant varieties of *S. typhimurium*, each unable to make certain amino acids and, therefore, crippled unless those nutrients were supplied to them. When they mixed a billion of each kind with bacteriophage, then spread them on a petri dish that contained no amino acids at all, a few colonies grew nevertheless. The bacteria acted as if their genes had been pooled. The same thing happened when one strain was resistant to the antibiotic streptomycin: the antibiotic resistance spread to the other strain. The phage, Zinder learned, packaged a piece of bacterial DNA into the virus particle instead of—or along with—its own DNA, and took it along when it infected the next cell. Zinder called this method of gene transfer "transduction."

A phage, like any virus, is not an organism in the usual sense. It is not a cell or collection of cells that extracts energy from chemicals to grow and move. A phage is a tiny packet of DNA wrapped in protein. Its protein coat protects it. It's also a DNA injector, whose purpose is to get the viral DNA from one cell to the next, where it can make a new crop of virus particles. Some bacteriophage look like lunar landing modules, their round heads full of DNA and connected by a stalk to spidery legs that attach to the bacterium. They are rather fantastic examples of self-organization: the whole process of viral assembly happens by itself. The proteins do what they are supposed to do by virtue of their structure. When enough of the DNA and of the structural proteins have been made inside the infected host cell, the head assembles itself and spools in a head-full of DNA.

Most of the time, only phage DNA is packaged. But every once in a while, a bit of bacterial DNA is pulled in and packed up. The machine just measures the length of DNA; it is blind to whether it is grabbing viral DNA or bacterial DNA. Released from the burst host cell, this transducing bacteriophage just as blindly does what all the other phage do: when it finds a new bacterial cell, it injects its DNA. But because its DNA is partly bacterial, it cannot make a full virus and so does not kill the cell. Instead the bacterial DNA finds its homolog in the new host and the bits of chromosomes recombine, just as in ordinary bacterial sex.

The ability of bacteriophage to assemble themselves from their protein subcomponents into a structure that can deliver and replicate its DNA became, in time, an essential part of the recombinant DNA toolkit. But before that, studies on another obscure problem led to the discovery of restriction enzymes—the molecular scissors with which Boyer cut the DNA he sent over to Cohen's lab to splice.

The problem was that seemingly identical bacteriophage sometimes grew on a particular strain of bacteria—and sometimes, mysteriously, they did not grow. This fact was rather irritating to bacteriologists, because it brought an element of unpredictability to phage experiments. The first step in clarifying the capriciousness of phage growth came in the 1950s. Whether or not a phage would reproduce depended on its history—on which strains of bacteria it had infected before. Swiss microbiologist Werner Arber, working at the University of Geneva with colleague Daisy Dussoix, found that the DNA of the infecting virus was either destroyed or replicated, depending on how the DNA had been modified when it grew in its previous host. If the new host modified DNA in the same way the old host had, the phage flourished. If not, the host was said to restrict the phage's ability to reproduce. These bacteria destroyed the invading viral DNA—cutting it to pieces with the enzymes that came to be called restriction endonucleases or restriction enzymes.

In 1978 Werner Arber shared a Nobel Prize with Hamilton Smith and Dan Nathans of Johns Hopkins University, who had independently isolated and used restriction enzymes to analyze viral DNA. These discoveries in turn inspired Paul Berg of Stanford, also sometimes called the father of genetic engineering, to combine chromosome bits from a virus and a bacterium. Although Berg did not clone this recombinant DNA in living cells, as Cohen and Boyer cloned their recombinant plasmids, he is nonetheless credited with creating the first recombinant DNA molecule. Berg's 1980 Nobel Prize states, "His pioneering experiment has resulted in the development of a new technology, often called genetic engineering or gene manipulation"—yet another birth of biotechnology.

More than 3,000 restriction enzymes are now known. They are among the most important tools in molecular biology today. Whole companies are devoted to the discovery, preparation, and marketing of them. Because there are many different kinds of bacteria, there are many different restriction-modification systems, each of which recognizes and cuts a very specific short sequence—between four and eight base pairs—of DNA. The precise number and location of the cuts made is unique to each DNA sequence.

DNA cut with restriction enzymes can be sorted by size using a technique called gel electrophoresis. The fragmented DNA is placed in depressions at one end of a tray of agarose gel, a stiff jelly made from seaweed extract. Electrodes are placed at both ends of the tray and a current is passed through the gel. While the current is on, the pieces of DNA migrate toward the positive electrode. The smaller the piece, the faster and farther it travels. The distinctive patterns of DNA fragment sizes, called restriction patterns, are the basis of DNA fingerprinting, the technology used today to match fathers with their children, to identify criminals or free the innocent, to distinguish a wolf from a dog or a corn plant from teosinte, to find the oldest flower or the most ancient breed of horse, or to find the gene responsible for an inherited disease. To start the process a scientist—or a detective—needs a sample of DNA.

Purifying a sample of DNA is simple; high school students routinely do it in biology classes, using spinach leaves or onions or wheat germ or kiwi fruit. It takes an hour or two and costs about $10 per student. First they grind the spinach or other plant with water and sand in a mortar and pestle. They add a little detergent (Palmolive, Dawn, or Woolite are all good) and some salt to break open the spinach cells. They scrape the spinach slurry into a large test tube and place it in a bath of hot water for 10 minutes. The slurry is iced, then filtered through four layers of cheesecloth. A little Adolf's Meat Tenderizer mixed into the filtrate destroys any proteins. Cold alcohol poured slowly down the side of the test tube makes the DNA clump together, forming a clear, gelatinous, mucus-like precipitate that will spool onto a glass rod twirled in the tube. This drop of pure DNA is then cut by molecular scissors: teachers can buy 2,000 units of a common restriction enzyme for $26.95, and the reaction itself takes 30 minutes in a warm water bath.

But to get a clear restriction pattern—or to find a specific gene, say a gene that encodes an enzyme for making a beta-carotene precursor that you might want to insert into a rice plant—the DNA fragment carrying the gene has to be identified, then amplified until there are enough copies to work with. While a drop of phage DNA gives a nice, discrete pattern of DNA bands, the same amount of DNA from spinach simply makes a fuzzy smear. Phage have just a few genes—no more than a hundred or so. One drop of phage DNA contains many copies of the phage's tiny genome. The restriction enzyme cuts each copy in the same few places, leaving lots of fragments with the same length. Because fragments of the same size end up in a single band, the result is a sharp, easy-to-read pattern. Plants and animals, however, have tens of thousands of genes. A drop of plant DNA contains many fewer copies of a much longer genome. The restriction enzyme makes cuts at many places, leaving thousands of fragments of various sizes. There are so many bands, each overlapping the next, that no pattern can be discerned, just a smear. It's the difference between a choir singing in four-part harmony and a cacophonous shouting crowd.

Plasmids and phage are ways to reduce the cacophony by amplifying one fragment at a time. Phage inject their DNA into bacterial cells, where it is cloned by a combination of enzymes coded for by the phage and by the bacterium. Hundreds or even thousands of exact copies of the bacteriophage DNA are made inside each bacterial cell. The cells fill up and eventually burst, releasing the particles to infect more bacterial cells.

Cutting phage DNA with restriction enzymes makes it easy to splice in a different kind of DNA, such as a bit of spinach DNA. Many restriction enzymes cut DNA in a way that leaves "sticky" ends. Rather than cutting the sequence AGTCAG, for instance, neatly in the middle, between T and C, it cuts after the initial A on one strand, leaving GTCAG dangling. It cuts the other strand with the same asymmetry, so that the middle four bases on each strand protrude. If spinach DNA is cut with the same restriction enzyme, its many, many fragments all have the same sticky ends. The stickiness comes from the same hydro-

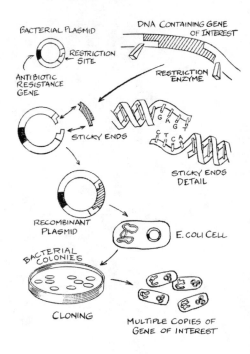

*How to clone a gene
in a plasmid*

gen bonding between complementary bases (A with T, and G with C) that holds the strands of the DNA double helix together.

When the two kinds of cut DNA, phage and spinach, are mixed, the cut ends stick together because the protruding single-stranded sequences are complementary. By getting rid of all but the most essential genes, a phage genome can accept a length of foreign DNA some 20,000 to 40,000 base pairs long—a decent stretch of DNA to analyze and one containing several to many genes, depending on the length of each gene.

One more enzyme is needed: a ligase, whose name, like ligature, comes from the Latin "to tie up or bind." If restriction enzymes are molecular scissors, DNA ligase is the molecular glue. It seals the DNA where the sticky ends overlap, making a continuous double helix again. Enough ligase for a class of biology students to clone spinach DNA costs $53.80. The recombined, or recombinant, DNA is then mixed with partly assembled phage particles extracted from bacteria. The phage complete their self-assembly process in the test tube. When added to bacterial cells, they inject and replicate their DNA, cloning the bit of spinach DNA they have picked up along with their own.

Plasmids, as Cohen and Boyer showed, are equally good vectors for cloning, especially plasmids with mutations that raise the number of copies from one plasmid per bacterium to hundreds. Recombinant plasmids are made in much the same way as recombinant phage DNA, by cutting with a restriction enzyme and sealing with DNA ligase, though to get the new plasmid into its host takes a third enzyme to breach the bacterial membrane—lysozyme, from tears, the same enzyme Alexander Fleming was studying when he discovered penicillin. (If lysozyme is not available, a brief electrical pulse or a 90-second burst of heat also works.) Then the bacterium does the cloning as it divides and grows. It treats the plasmid just like its own DNA, copying it carefully and distributing it to its daughter cells.

In the experiment for high school students, the recombinant DNA cloned in the bacteria contains the whole genome—all the DNA—of spinach. Although each individual bacterium and its progeny have cloned just one spinach DNA segment, even a small test tube of bacteria contains millions of different clones. But scientists generally want to single out and clone just one gene. To do so, the bacteria-filled broth is first diluted, then spread very thinly over the surface of an agar layer in a petri dish. The solid agar both nourishes the bacteria and confines them, letting a single cell—and all the cells that accumulate as it divides—stay in the same place, forming a colony. The plasmid used as the cloning vector carries a marker gene—one that makes the host cell resistant to an antibiotic, for instance. The antibiotic-resistance marker gene identifies the bacteria that picked up the plasmid. When the agar is laced with antibiotic, any bacterium without a plasmid dies. Each plasmid-containing bacterium grows into a colony of cells, each containing copies of the same antibiotic-resistant plasmid, each with a copy of the same spinach DNA fragment.

Phage clones are prepared in a similar way: the recombinant phage preparation is diluted, mixed with bacteria, and plated on solid agar. When the first crop of phage particles is released, it can infect the bacteria only in its immediate vicinity. The result is that the lawn of bacteria that grows overnight in a warm incubator is riddled with clear spots. These are areas where the bacterial cells have burst and released their crop of phage particles in place. Each spot or plaque contains the progeny of just one original phage particle. The particles in each plaque contain just one of the original fragments of DNA.

In the early days of cloning, the next step—identifying the right gene from among this collection, this library of clones, each growing in its own bacterial colony—was the most time-consuming. But in 1983 Kary Mullis had an idea that would shorten the time to clone a gene from months to hours. "It was a chemical procedure," Mullis wrote, "that would make the structures of the molecules of our genes as easy to see as billboards in the desert and as easy to manipulate as Tinkertoys." Mullis won the 1993 Nobel Prize in chemistry for the gene-copying technology he called the polymerase chain reaction, or PCR. Writing about the emergence of biotechnology, anthropologist Paul Rabinow explained: "PCR has profoundly transformed the practices and potential of molecular biology. It makes abundant what was once scarce—the genetic material required for experimentation. Not only is this material abundant, it is no longer embedded in a living system." You don't need a bag of spinach to have enough DNA to clone. A sufficient amount can come from a single leaf—or from a spot of dried blood, a swipe with a cotton swab on the inside of your cheek, a wolf hair, or a bit of corn pollen.

The Cetus Corporation of San Francisco, where Mullis worked as a technician, was one of the first biotechnology companies. Founded in 1971, it made therapeutic proteins, like interleukin II, vitamins, and vaccines. It produced genetically engineered yeast for making alcohol, and tried to improve the yield of the bacterial fermentations that produced antibiotics. Mullis was hired in 1979 to make short strings of DNA called oligonucleotides.

At first he liked his work. "It was really fun to learn how to synthesize DNA," he told the Smithsonian Institution in 1992. "It was just organic synthesis, pure and simple." But he soon grew bored. A surfer who was not shy about the fact that he liked pictures of naked women or that he had used LSD, Mullis was known for wild ideas, according to his co-workers at Cetus. He was abrasive, combative, surly, and he had an irreverent, even belligerent, attitude toward established procedures. He was always tinkering, trying to make a process faster or more efficient.

Making oligonucleotides was repetitious. Mullis wrote, "Most people didn't like to do things over and over, me in particular. If I had to do a calculation twice, I preferred to write a [computer] program instead." Rather than quit his job, he began to tinker. As he tells the story in his autobiography, *Dancing Naked in the Mind Fields*, the idea for PCR came to him one night. What he imagined was a loop—the kind programmers use to tell a computer to perform the same task over and over again. He would set up a chain reaction—a chemical reaction that fed on itself, cycling over and over again—based on the action of DNA polymerase, an enzyme whose function is to copy DNA.

First create two short strips of DNA, two oligonucleotides, each 15

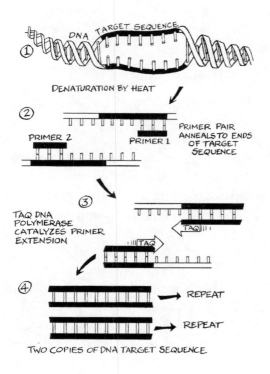

The polymerase chain reaction, or PCR

or 20 nucleotides long. Mix in a sample of DNA from a spinach or human or any other source. Heat the sample so that the DNA helix comes apart. As the mixture cools, the two oligonucleotides, or primers, will quickly seek out and stick to their complements on the longer strands before the original DNA can zip back up. If the test tube also contains a DNA polymerase enzyme and a collection of free nucleotides, then the polymerase will start at one primer and build a new DNA chain until it reaches the second primer. Reheat the sample so that the new double strand separates, and repeat. The chain reaction can create a hundred billion copies in an afternoon.

At first, Mullis relates, "however shocking to me, not one of my friends or colleagues would get excited over the potential for such a process. True, I was always having wild ideas, and this one maybe looked no different than last week's. But it *was* different." Cetus nearly dropped the idea (it later sold PCR for $300 million). It took almost two years for the scientists at Cetus to get Mullis's idea to work. Only when they tried using an enzyme that could handle the heating cycles without being deactivated—an enzyme called *Taq* polymerase, extracted from a bacterium that grows in a hot spring in Yellowstone Park—did the result seem worth the effort. As one of the scientists on the project said, "The holy grail had been achieved. It had worked better than even we had fantasized."

Today every molecular biology lab has an automated PCR machine. As Mullis puts it, "PCR is to DNA what the screwdriver is to screws." The procedure is simple enough that it can be done in a kitchen, according to the *Scientific American* column "The Amateur Scientist." Although it is "quite challenging" to do properly, the equipment is minimal: a few beakers, a blender, candy thermometers for the hot-water baths, cotton swabs, plastic coffee stirrers, latex gloves, and, says Mullis, "about 20 different things you have to measure out, each of them dependent on all the others." In 2000 the Society for Amateur Scientists sold a PCR kit for $40 (their gel electrophoresis kit was $60).

If you know the DNA sequence of the gene, it is simple to make two short primers (the 15 to 20 nucleotides at the start and the end of the gene), run PCR and have a test-tube-full in a few hours. Such a PCR-amplified DNA fragment can be spliced directly into a plasmid or used in other ways. If the gene's sequence isn't known, it is harder to clone the gene, but not impossible. Working backward from the protein, a primer can be designed to find and amplify the gene. Because a protein is a string of amino acids, each of which is coded for by a codon made up of three nucleotides, the sequence of codons can be translated back to give the gene's nucleotide sequence. The difficulty comes from the fact that the amino acid code is degenerate: there are 20 amino acids, but 64 codons. Like synonyms, several codons can signify the same amino acid. To get around this complication, a mixed primer is necessary, one that consists of all of the possible nucleotide sequences—using all the possible synonyms—that could code for a particular string of amino acids.

To verify that it is the correct gene, a PCR-amplified gene must be sequenced—a step that was automated with the invention of the DNA sequencing machine in 1986. Today DNA sequencing is routinely done on a huge scale: In addition to the entire human genome sequence, many other genome sequences have been deciphered, including those of mice, rats, the fruit fly *Drosophila,* the nematode *C. elegans,* the yeast *Saccharomyces cerevisiae,* the weed *Arabidopsis,* two varieties of rice, and those of many, many bacteria. Whole genome sequences can now be accessed over the Internet.

Adding genes to plants required all the inventions and techniques just described. But to put a daffodil gene into rice also required the help of a natural genetic engineer, *Agrobacterium tumefasciens.* Only when this common soil bacterium was understood—and harnessed—did the genetic engineering of plants become a reality.

A. tumefasciens is closely related to *Rhizobium meliloti,* the bacterium that helps legumes like alfalfa and peas fix nitrogen. But rather than being helpful like its relative, *Agrobacterium* was known by 1907 to be an agricultural pest. It causes galls to form on fruit trees, walnuts, grapes, roses, and other valuable plants. Such tumors grow in places where the plant has been wounded. The cells in the gall produce chemicals that the bacteria use for food, so that *Agrobacterium* can be said to farm the plant without killing it. "For a long time, perhaps millions of years, the common soil bacterium *Agrobacterium tumefasciens* has been doing what molecular biologists are now striving to do," wrote Mary-Dell Chilton of the University of Washington in 1983. "It has been inserting foreign genes into plants and getting the plants to express those genes in the form of proteins."

That it was inserting *genes*—DNA—was not understood until the late 1970s. Thirty years before, Armin Braun of the Rockefeller Institute for Medical Research showed that *Agrobacterium* injected a tumor-inducing principle that transformed an ordinary plant cell into a constantly growing tumor cell. By analogy to the famous experiment done in 1928 by Fred Griffith, who showed that a deadly strain of the pneumonia bacterium could pass its virulence principle to a previously harmless one, French scientist Georges Morel predicted that this tumor-inducing principle of *Agrobacterium tumefasciens* would be DNA. But for a bacterium to transfer DNA to a plant—breaking the barrier not only between species, but between kingdoms—was almost unimaginable in the 1950s and 1960s.

In 1974 Jef Schell and Marc Van Montagu at the University of Ghent in Belgium discovered a large plasmid in the virulent strains of *Agrobacterium* that was missing in the harmless strains. The pest's method of operation began to become clear. A year later Mary-Dell Chilton, Milton Gordon, and Eugene Nester of the University of Washington confirmed that the plasmid was the tumor-inducing principle. *Agrobacterium* inserts a piece of the large plasmid into the plant's genome and tricks the plant into expressing it, manufacturing proteins that benefit the bacterium, but not the plant. The method it uses to transfer its genes, though, more closely resembles bacterial sex: *Agrobacterium* constructs a mating tube through which the plasmid travels.

When the complete genome of *Agrobacterium tumefaciens* was sequenced, it was found to have 5.67 million base-pairs of DNA, an estimated 5,400 genes, on 4 structures: a circular chromosome, a linear chromosome, and 2 smaller circular plasmids. One of these plasmids is the genetic engineer. A short stretch of fewer than 80 genes, a region called *vir* for "virulence," is the critical part. It holds 7 groups of genes, or operons, labelled *A* through *G*, each containing from 1 to 11 genes. The first operon, *virA*, encodes a protein that sits on the bacterium's cell membrane. It senses the chemicals released when a plant cell is injured, and immediately signals *virG*, whose job it is to turn on the other genes in the *vir* region. The *virB* proteins create the connecting tube. A protein encoded by *virD* nicks the plasmid's DNA strand in two particular spots, releasing a long bit of single-stranded DNA known as the Transfer-DNA, or T-DNA. The *virE* proteins then coat the T-DNA and usher it through the mating tube and into the plant cell and thence to its nucleus. There the T-DNA seeks out and links up with the plant cell's nuclear DNA, permanently transforming the plant cell into a tumor cell.

On the single-stranded stretch that is transferred, Chilton and her colleagues found a transposon. It was "a sizable piece of DNA," she wrote. "Its presence in the tumor cells meant that a large segment of foreign DNA, enough to carry several genes, can be transferred along with the T-DNA."

Four laboratories began working immediately on the idea of using *Agrobacterium* as a vector: Schell and Van Montagu at the University of Ghent in Belgium, Eugene Nester at the University of Washington in Seattle, Mary-Dell Chilton at Washington University in St. Louis, and Rob Horsch and his colleagues at Monsanto. They announced their success simultaneously in 1983. The goal of these experiments had been to see if the *Agrobacterium* plasmid could be used to introduce a gene— any gene—into a plant. To show that it could be done, a gene that makes bacteria resistant to the antibiotic kanamycin was substituted for the tumor-forming genes. Kanamycin kills plant cells—it interferes with their ability to make proteins. Without the new gene, all of the plant cells in the experiment would die if they were exposed to the antibiotic. The techniques used to construct the recombinant *Agrobacterium* plasmid were similar to those that Cohen and Boyer used (although a bit more complicated because the *Agrobacterium* plasmid was so very large).

Bits of tobacco leaves were then dipped into a suspension of *Agrobacterium* containing the recombinant plasmid carrying the kanamycin-resistance gene. The infected leaf bits were put on agar that contained growth hormones, which allowed the cells to grow into a disorganized callus. The agar medium also contained kanamycin, which prevented the growth of cells that lacked the recombinant plasmid. Gradually the callus cultures were weaned off growth hormones and began to produce shoots, then roots. The plants not only grew on the antibiotic, the DNA that codes for kanamycin-resistance was now linked to tobacco DNA—and plant genetic engineering was born.

When the idea of Golden Rice surfaced, the *Agrobacterium* technique was thought to work only in broad-leaved plants, not in grasses. The technique was assumed to be useless for engineering rice, wheat, or corn so Ingo Potrykus ignored it, continuing to insert naked DNA into protoplasts. But in time it was found that *Agrobacterium* could transfer genes to grasses, and it was by using *Agrobacterium* that Xudong Ye eventually transferred the beta-carotene-producing daffodil genes into rice in Potrykus's lab.

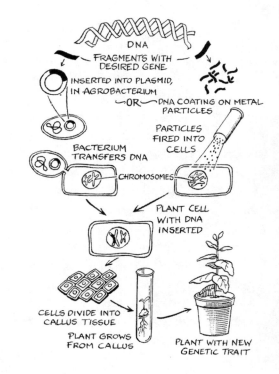

Two ways to give a plant a new gene: the Agrobacterium *method and biolistics*

The use of naked DNA succeeded when John Sanford at Cornell University took what's been called the "cowboy approach." He used a modified BB gun to shoot tiny gold or tungsten particles coated with DNA through the intact walls of plant cells. The small holes healed and, in some of the cells, the DNA became part of the plant's chromosomes. These cells could then regenerate into whole plants.

But regardless of the technique that gets the DNA into the plant, the result is a DNA chimera: a piece of foreign DNA is spliced into the plant's genome. As Boyer and Cohen noted when they coined the term in 1973, a DNA chimera is no different than the chimeras that have been created for the last 3,000 or so years by grafting. It is as natural—or as artificial—as an apple tree. Neither would exist, in their current forms, without human intervention. They differ only in scale. The genetically engineered plant is created using molecular scissors—restriction enzymes—not a knife or scalpel. What's grafted on is a few well-studied genes, not a bud or a branch. Each technique combines genes from different species. Even crossing kingdoms to put a bacterial gene into a plant is not new: *Agrobacterium* has done it for millennia. What's new is that the pool of possible genes from which plant breeders can now choose has grown very much larger.

7

THE PRODUCT OR THE PROCESS

Select any time point in the history of agriculture—say, agriculture and agricultural research in 1990—and we cannot really say that before this point things were still done in a natural way. Go back another 10 or 20 years, and we still could not maintain that up to that point, plant breeders were doing no more than nature itself could have managed. Why, in the debate on natural versus unnatural, should we draw the line right here, right now, at the point where genetic engineering has entered the scene?

—Per Pinstrup-Andersen and
Ebbe Schioler (2000)

"What has long appeared to be simply the agent of a bothersome plant disease," wrote Mary-Dell Chilton in 1983, "is likely to become a major tool for the genetic manipulation of plants: for putting new genes into plants and thereby giving rise to new varieties with desired traits." At the time, coaxing *Agrobacterium tumefasciens* into adding a new gene, any gene, to a plant, was itself bothersome. Building the vector, the ring of DNA that the bacterium would insert into the plant cell, was a laborious process, involving several painstaking steps. Still, the possibilities excited Roger Beachy, a scientist then in the same department as Chilton at Washington University in St. Louis. "While she was making her vectors," he recalled in 2003, "I was imagining what I could use them for."

"Vector" comes from the Latin for "carrier." Biologists had commonly used it to describe the carrier of a disease—the tsetse fly is the vector for sleeping sickness; the aphid is the vector for tomato mosaic virus. To Beachy, who had been studying plant viruses since graduate school, Chilton's new vectors were an opportunity to turn the meaning of the word on its head. He was investigating the phenomenon of cross-protection. "If you infect a tomato plant with a mild strain of tomato mosaic virus, you can protect it from infection by a severe strain," he explained. "This had been known for more than 20 years." Why it worked, though, was a mystery. It was like vaccination, except that plants don't have an immune system. Beachy wondered what part, precisely, of the virus provided the protective effect—and could he make Chilton's vector carry that vaccine-like bit into a plant cell?

Tomato mosaic virus, like many plant viruses, is a single-stranded slip of RNA coated with protein. Once inside a cell (placed there usually by an aphid as the insect tastes a leaf), the protein coat unwraps and the naked RNA begins interfering with the cell's protein-making machinery, directing it to make viral proteins instead of its usual plant proteins. Beachy thought the key to cross-protection might be the coat protein.

By 1981 scientists had sequenced the tobacco mosaic virus genome, a close cousin to the tomato mosaic virus. It was a manageable 6,400 nucleotides long. Beachy had earlier mapped the order of this virus's genes, so he knew more or less where to look for its coat protein sequence. He made DNA clones of it and of other parts of the genome, then approached Chilton with the idea of using her vectors to insert each bit of viral DNA into a plant cell. If the plant carrying the coat protein gene became resistant to the virus and the others did not, he would know the source of cross-protection.

Chilton was game to try, but an offer to work for Ciba-Geigy came before the experiment could be done. Rather than giving up, Beachy sought out one of Chilton's competitors in the so-called *Agrobacterium* race. Across town in St. Louis was the headquarters of Monsanto. There, in a laboratory for which both Chilton and Jef Schell from the University of Ghent had consulted, Robert Fraley and Stephen Rogers agreed to take on the project. With Fraley's and Rogers's assistance, and with funding from Monsanto, Beachy and a student made a vector. To the coat protein gene they attached a promoter, a DNA sequence that turns on a gene. A promoter is critical: without it the process by which DNA is transcribed into RNA and translated into protein stops at step one. The promoter acts like a hook: to start transcribing the gene, the enzyme RNA polymerase latches on to the promoter.

Beachy chose the wrong promoter. Although *Agrobacterium* neatly carried the new gene into the plant cell, where it combined with the plant's own DNA in a process that came to be known as transformation, the promoter was too weak to be effective, Beachy said. "We could barely detect the coat protein. It was there, but at extremely low levels. The student was terribly disappointed."

Fraley and Rogers, however, knew a trade secret: Monsanto had a new promoter. Designated 35S for the length of the original RNA whose synthesis it directed, it came from cauliflower mosaic virus, known as CaMV. Patented in 1994 by Fraley, Rogers, and Rob Horsch, all working at Monsanto, the CaMV 35S promoter became the standard gene promoter used in experiments with plants. In the early 1980s, Beachy said, "They wouldn't tell me much about it. Monsanto did the transformation and we got the plants back. Our first success came with tobacco. Then came tomato, then petunia. We had a lot more coat protein, so we had resistance." Expressing just the gene coding for the virus's protein coat was indeed enough to protect the plant against infection.

As with many advances, including smallpox vaccinations in people, the knowledge *that* it worked preceded a deeper understanding of *how* it worked. Several years later scientists working in Beachy's lab discovered that, in plants containing the coat protein gene, the virus was unable to come apart. The viral RNA was not released into the cell, and the cell was not infected. This coat protein-mediated resistance, as they called it, allowed plants to fight off infection by more than one closely related virus. "For example, plants that were made resistant to tobacco mosaic virus," Beachy explained, "were also resistant to tomato mosaic virus."

It was discovered recently that virus resistance can also develop by a different biochemical mechanism called post-transcriptional gene silencing. When a gene—whether viral or some other—is expressed at too high a level, a feedback process is triggered that shuts down its function. This process affects the RNA transcript. The gene continues to be transcribed, but the transcript is destroyed. With no template from which to make the protein, no protein is made. A gene has to be almost identical to the overexpressed gene to be shut down along with it. No other gene in the plant is affected.

Post-transcriptional gene silencing works directly on RNA. Most plant viruses use RNA as their genetic material; their genes are made of RNA, not DNA. When a virus infects a plant cell, it directs the plant cell to make lots of viral RNA, triggering the gene silencing mechanism directly. Post-transcriptional gene silencing has a memory, as does our immune system, though it works in a very different way. When a plant is infected a second time with the same virus or a very closely related virus, the gene silencing system is reactivated, and the invading viral RNA is destroyed as soon as it takes off its coat.

This phenomenon is the cross-protection Beachy was interested in. At the time it seemed logical—perhaps by analogy to the immune system—to suppose that cross-protection worked because the plant recognized the viral coat protein. In some cases, the coat protein is indeed the active agent. But more often this type of gene silencing works by recognizing and destroying the RNA. RNA silencing is as precise as the human immune system, although it works by a completely different principle.

Beachy, Fraley, and Rogers's successful T-DNA contained the DNA code for the coat protein. It was inserted into the plant's genome using *Agrobacterium.* It worked because the strong CaMV 35S promoter caused the gene to be transcribed so often, that is, to produce so much RNA, that the post-transcriptional gene silencing system was triggered. The RNA of an infecting virus was thus recognized and destroyed as soon as it entered the cell.

In 1985 Beachy, Fraley, and Rogers applied for a patent on their technique to protect crops from viral diseases. The patent was finally awarded in 2003. By then, virus-resistant varieties of squash, papaya, tomato, pepper, cucumber, sugar beets, and plums—all created using the coat protein technique—were on the market. Virus-resistant sweet potatoes were in development. And Beachy, by then director of the Donald Danforth Plant Science Center, a private research institute in St. Louis, was at work on virus-resistant rice.

Some of the virus-resistant plants, as well as the technique used to make them, are patented. Patents are widely misunderstood as ownership of an idea or an invention. Yet to get a patent the inventor must make public enough information that someone "skilled in the art" can reproduce the invention (if given a license or other legal right to do so by the patent owner). A patent merely lets the inventor prevent others from using the idea without permission for a limited period of time, usually 17 to 20 years. It is easier to own, to control, an idea if it is kept secret and not patented.

Patents are meant to encourage creativity by letting the inventor get a head start on making a profit. But they are also intended to further scientific progress. Before Monsanto patented the CaMV 35S promoter it was a closely held trade secret. Even a collaborator like Beachy was kept in the dark. Once patented, it was available to researchers around the world, sometimes free, sometimes for a licensing fee, depending on whether the researcher was at a university or other research institute or in a company.

The patent system has been affecting western agriculture since at least 1750. Until the mid-1700s farmers in Europe and America were most influenced by the Latin poet Virgil. According to historian Mauro Ambrosoli, who studied farming handbooks published between 1350 and 1850, much of the advice given was not based on science or on practical experience, but on classical poetry. And much of it was particularly absurd when applied to England: Virgil wrote 2,000 years ago, for instance, "If fat the soil, let sturdy bulls upturn it from the year's first opening months, and let the clods lie bare till baked to dust by the ripe suns of summer."

Jethro Tull, a lawyer and Oxford graduate, challenged this status quo. Experimenting on his own Howberry Farm, he advised reducing the quantity of seed sown. Instead of broadcasting seed—flinging it out of baskets by the handful—he invented a seed drill to plant individual seeds neatly in rows. Instead of sowing rye, barley, and clover together, to keep down the weeds, he devised a horse-drawn hoe to handle the weeding. He saved seeds from his own harvests to use the next year. For this he was fiercely attacked by the Private Society of

Husbandmen and Planters, whose leader was gardener and seed merchant Stephen Switzer, the author of "good works on agronomy," Ambrosoli writes, that were "frankly inspired by" Virgil. The seeds sown in England were, at the time, often grown in Scotland and imported for resale by seed merchants like Switzer. Buying new seed each year prevented blight and smut and cut down on weeds; seeds from the cold north also gave a better yield when planted in the fertile south. "The attack on Tull becomes more understandable," Ambrosoli continues, "if connected with the wish to maintain control over the seed market and to avoid losing buyers of the illustrated booklets sold with the seeds."

Jethro Tull was a gentleman farmer. He did not patent his seed drill or his horse hoe, but after 1750, Ambrosoli notes, "every new invention concerning agricultural machinery was jealously defended with patents." In America by 1836 the head of the U.S. Patent Office, Henry Ellsworth, believed it was his responsibility to encourage the introduction of new plant varieties along with the invention of new machines. The idea that plants themselves could be patented grew stronger in the 1870s, when Louis Pasteur, the inventor of milk pasteurization, attempted to patent a type of yeast. Luther Burbank was among those who argued at the time that if a yeast could be patented so could a plum or a potato. Yet, as his biographer, Peter Dreyer, notes, "The degree to which cultivated plants were human creations was not generally recognized in those days." No one thought of a potato as intellectual property.

It was Burbank's posthumous testimony that led to the passage of the first bill in the United States to include plant breeders among the ranks of inventors. As Dreyer writes in *A Gardener Touched with Genius*, one of the main opponents of the bill was Congressman Fiorello La Guardia, later mayor of New York. "Acting apparently on misunderstood protests he had received, La Guardia successfully blocked passage, until the sponsoring congressman, Fred S. Purnell of Indiana, inquired what he thought of Luther Burbank. 'I think he is one of the greatest Americans that ever lived,' the New Yorker replied flamboyantly. Purnell then proceeded to read into the record a letter written by Burbank to Paul Stark shortly before his death. 'I have been for years in correspondence with leading breeders, nurserymen, and federal officials, and I despair of anything being done at present to secure to the plant breeder any adequate returns for his enormous outlays of energy and money,'" Burbank had written. The letter went on:

> A man can patent a mouse trap or copyright a nasty song, but if he gives to the world a new fruit that will add millions to the value of earth's annual harvests he will be fortunate if he is rewarded by so much as having his name connected with the result. Though the surface of plant experimentation has thus far been only scratched and there is so much immeasurably important work waiting to be done in this line I would hesitate to advise a young man, no matter how gifted or devoted, to adopt plant breeding as a life work until America takes some action to protect his unquestioned rights to some benefit from his achievements.

As Dreyer relates, LaGuardia immediately withdrew his objection to the bill and it passed both House and Senate.

The Plant Protection Act of 1930 covered cultivated plant varieties, whether they had been developed by a breeder like Burbank, or discovered in the field by a plant explorer, as long as they were reproduced asexually. (Curiously, tubers were excluded: Burbank still could not have patented his famous Idaho potato.) Such a plant patent was intended to prevent anyone except the developer or discoverer from producing the variety for a number of years. The stricture that the plant had to be propagated asexually—through grafting, cuttings, runners, or the dividing of bulbs—arose from the need to preserve the distinctive characteristics of the variety, which sexual reproduction tended to disrupt.

Yet many sexually reproducing plant varieties do maintain their distinctiveness—growers say they breed true—and by the 1960s these were given patent-like protection in some European countries. After a number of unsuccessful attempts, the U.S. Congress did so as well, passing the Plant Variety Protection Act in 1970. To receive a plant patent, the plant variety must be uniform, stable, and distinct from other varieties. To encourage breeders to develop new varieties, the act set up a mechanism to provide exclusive marketing rights by certifying seed.

The 1970 act excluded fungi and bacteria. So when Ananda Chakrabarty, through his employer, General Electric, applied for a patent on a strain of bacteria that could efficiently degrade crude oil, his application was denied. Crude oil is a complex mixture of hydrocarbons. Several strains of the bacterium *Pseudomonas putida* were known to produce enzymes that degraded one or another of those hydrocarbons—that is, the bacteria ate the hydrocarbons, nourishing themselves and reproducing. But mixtures of strains that could eat different hydrocarbons weren't very efficient when faced with the task of degrading crude oil. Some grew better than others. The overall efficiency of converting crude oil to bacterial biomass (which is 75 to 80 percent protein) wasn't very good.

Chakrabarty discovered that the ability to degrade hydrocarbons was due to genes on plasmids that the bacterium carried. Different strains had different plasmids and so could degrade different hydrocarbons. Chakrabarty devised a way of combining the multiple plasmids in a single strain. Usually, similar plasmids can't replicate in the same bacterium—they exclude each other—so Chakrabarty used ultraviolet radiation to fuse the multiple plasmids inside of bacterial cells, creating a strain with one large plasmid. It carried genes coding for enzymes that could degrade many different hydrocarbons, and it grew better on crude oil than did a mixture of strains, each carrying a single plasmid. Because Chakrabarty's bacteria could be useful in cleaning up oil spills, General Electric decided to apply for a patent on the process of constructing bacteria with such fused plasmids, as well as on the organism itself.

The patent application was filed in June 1972. In late 1973 the U.S. Patent and Trademark Office issued its decision. It accepted the process claim, but did not grant a patent on the bacterium, which it said was a product of nature and therefore not patentable. General Electric's patent attorney appealed the decision, arguing that Chakrabarty's bacterium was not a product of nature; it had been changed by fusing incompatible plasmids. The appeal board acknowledged this argument, but still rejected the patent claim because it sought to patent a living organism.

General Electric appealed to the U.S. Court of Custom and Patent Appeals. In 1978 the court granted the patent. In its view, as long as a microorganism was useful, novel, and the product of human intervention, it merited patent protection. The decision that living organisms could be patented was appealed to the Supreme Court, which returned it to the Court of Custom and Patent Appeals, where it was reaffirmed. The court argued that patent law should apply to "anything under the sun that is made by man," and that this included living organisms.

The Chakrabarty case changed how patent law was interpreted. Although Chakrabarty didn't use molecular techniques in constructing his oil-eating bacteria, by the time the decision was final, most of the recombinant DNA techniques in use today had been invented. The first biotechnology companies had been started, and the first patent applications had been filed.

Based on the Chakrabarty precedent, the patent office board of appeals issued an administrative ruling in 1985 that plants could be patented without following the special provisions of the 1930 Plant Patent Act or the 1970 Plant Variety Protection Act. In 1987 the board ruled that a polyploid oyster, a multicellular animal, could be patented. In issuing this ruling, Donald Quigg, the assistant secretary and commissioner of patents and trademarks, explained that the patentability of living organisms required that "they must be given a new form, quality, properties, or combinations not present in the original article existing in nature."

One consequence of patenting "anything under the sun that is made by man" is that nearly any DNA sequence can be patented. A cloned DNA sequence—although identical to the sequence in the organism from which it was derived—is not considered to be a "product of nature" because it required human intervention to clone it. Patents have therefore been issued on isolated bits of DNA. This practice is consistent with the longstanding one of issuing patents for chemical compounds purified from naturally occurring mixtures, but it continues to be problematic. For plant breeding, the patenting of DNA has two faces. On the positive side, it brings such trade secrets as the CaMV 35S promoter out into the open. On the negative side, it creates an additional, time-consuming step between discovery and application.

As soon as his patent application was filed in 1985 and he was free to talk about his invention, Roger Beachy called a colleague, Jerry Slightom at UpJohn Company, which then owned Asgrow Seeds. In a few months Asgrow had negotiated the rights to the viral coat protein technology and had begun developing a virus-resistant variety of yellow squash. Later, Asgrow was acquired by a multinational seed company, and a broad technology licensing agreement that included Beachy's invention was made with Monsanto. One item in that package concerned soybeans. "The access to elite, high-yielding soybean lines was an opportunity," Beachy recalled, "for a series of experiments that—10 years later—would become Roundup Ready soybeans."

Slightom, however, did not stop with squash. He placed a call to Cornell University plant pathologist Dennis Gonsalves, an expert on the papaya ringspot virus and a native of Hawaii, and told him about Roger Beachy's new technique. Gonsalves talked with Beachy, then approached John Sanford, also at Cornell. Sanford had by then applied for a patent on his gene gun (the so-called "cowboy method" of transforming a plant), and was in the process of forming a company, Biolistics Inc. (from biology and ballistics), to market the new tool. He was close to announcing the first transformation of corn. Gonsalves persuaded him to experiment with papayas and the coat protein of the papaya ringspot virus as well.

The papaya ringspot virus is spread by aphids from weeds and other wild plants that generally show no symptoms. Papaya are grown from seed, but don't begin producing for 9 to 12 months. A fully grown, 30-foot-tall papaya tree is productive for at least 3 years. When infected by papaya ringspot virus, however, the plants are stunted and their fruit is misshapen and tasteless; eventually the papaya tree dies, depriving the farmer of not only one harvest, but several.

In the United States, the Hawaiian islands are the only place papayas will grow. Oahu had been the center of papaya growing until the virus wiped out its papaya farms in the 1950s. Papaya growing moved 19 miles across the water to the Puna district of Hawaii, where the virus was kept under control by burning all infected papaya plants and all wild plants or other crops nearby that could harbor the virus. By the 1990s papaya was being grown by 200 to 300 farmers, each averaging

10 acres of trees; their annual production of over 50 million pounds of fruit was valued at $17 million.

Yet the papaya ringspot virus was only stalled, not defeated. Nineteen miles of ocean is not an uncrossable distance for an aphid. Aphids have traversed the Pyrenees from France into Spain, and the Atlantic Ocean from Europe to North America. The easiest (and most probable) way for them to travel is to hitch a ride on a fruit or twig or flower carried by a person. And once on Hawaii the virus would spread very rapidly, because the island's climate is very comfortable for aphids. The winters are so mild, the insects do not need to mate and lay eggs to keep the population going through the colder months, and so there are no male aphids in Hawaii. Instead every day each female aphid gives birth to between 8 and 22 live clones of herself. If she carries the papaya ringspot virus, so do they. The population of virus-carrying aphids quickly reaches the millions.

Insecticides provide no relief. It takes an aphid less than a minute to infect a plant. By the time a farmer has noticed the infestation and gotten out the spray, the damage is done. Millions of dollars had been spent on conventional breeding programs to create a papaya that could withstand the ringspot virus, but with little success. No naturally occurring genes for resistance to papaya ringspot virus had been found in any papaya variety.

In 1987 Gonsalves recruited horticulturalist Richard Manshardt at the University of Hawaii and plant physiologist Maureen Fitch from the USDA's Hawaii Agricultural Research Center; later University of Hawaii plant pathologist Steve Ferreira would join the group. Without knowing exactly how having the papaya ringspot virus's coat protein gene could protect the papaya from the virus, the team decided to try it. They cloned the coat protein gene from a Hawaiian isolate of the virus, then used Sanford's gene gun to insert a vector containing the gene into young embryos of a commercial Hawaiian papaya named Sunset. Using tissue culture techniques, they grew the transformed embryos into plants and tested their resistance to the virus. They found the first plant to show resistance, number 55-1, in 1991.

In 1992—coincidentally the year the virus was detected in the Puna district—number 55-1 underwent field trials in Hawaii. It proved to be remarkably resistant to the virus which, meanwhile, was spreading rapidly. By 1994 almost half of Puna's papaya acreage was infected, and a number of farmers had gone out of business.

The University of Hawaii researchers crossed 55-1 with commercial cultivars to produce the varieties SunUp and Rainbow. In 1994 the team received a permit from the USDA to conduct a large-scale field trial of these papayas near an abandoned orchard in Puna. The results were striking. Within a year, all the plants without the coat protein gene were infected, while only three of those with the new gene showed symptoms. Better yet, all of the resistant plants were still healthy two years later. The fruit was good, and the farmers approved.

But hurdles remained. Before the papaya seeds could be distributed to farmers—free of charge—licenses had to be obtained from the corporations and universities who held patents on the genes and processes that had been used to make the variety. Then the virus-resistant papaya had to be vetted by the appropriate government agencies. Hawaii's Papaya Administrative Committee (a group of papaya growers organized by the USDA) began negotiating for licenses from Monsanto, Asgrow Seeds, Cambia Biosystems, and the Massachusetts Institute of Technology. The University of Hawaii and Cornell scientists prepared the documents required by the government agencies. It wasn't at all clear that these two processes could be accomplished in time to save the papaya industry in Hawaii. Because the problem was so severe, the USDA funded a program to produce seeds of the resistant varieties so that they would be ready for immediate distribution—if and when the permissions came through.

~~~~~

In the United States any crop modified using molecular techniques—and only those crops—must be scrutinized by at least two, and often three, government agencies. The crop is evaluated as a potential toxic substance, pesticide, or plant pest, and sometimes as a food additive as well.

In view of humankind's long history of tinkering with food plants, this state of affairs is very odd. Over the last century breeders have learned to churn up plant genomes in many new ways. They cross plants from different species and even different genera. They use tissue culture, chemicals, and radiation to make mutants—plants that might be more resistant to drought, disease, and pests, or that might provide more or tastier food than the unmutated variety. Their techniques have become increasingly complex, and increasingly invasive.

For example, in 1979 Shivcharan S. Maan of Fargo, North Dakota, received United States Patent No. 4,143,486. "The invention," wrote Maan, "satisfies the long felt need for a relatively simple, commercially feasible method of producing hybrid wheat seeds." Maan's method begins with a single cell from the weed *Aegilops squarrosa*, or goat grass, one of the ancestors of modern wheat. First the nucleus of the cell is removed and discarded. Then a new nucleus, taken out of a cell of bread wheat, *Triticum aestivum*, is inserted into the goat grass cell. This alloplasmic wheat cell is grown, using tissue culture techniques, into a mature plant, and the seed is harvested. This seed is then exposed to radiation or soaked in a mutagenic chemical. In the example included in his patent, Maan notes that 500 seeds from plants having *Ae. squarrosa* cytoplasm and a *T. aestivum* nucleus were exposed to a mutagenic chemical for 16 hours before being planted in a greenhouse. The seed spikes of the mature plants were covered with plastic bags so that the plants would self-pollinate. "The 45,000 seeds obtained from self-pollination were harvested and planted in the field." Maan and his colleagues did not know what types of mutations the chemical treatment might have caused. To learn, they grew the plants to maturity and "visually examined" them for "abnormalities," choosing the 39 plants that could not make pollen to include in further breeding. The thousands of others they simply plowed under.

No federal agency checked to see if these plants were toxic or allergenic. No federal agency needed to be asked before the plants could leave the greenhouse. No federal agency needed to approve the scale-up of the field tests. Asked if he needed to file an environmental impact statement before field-testing his new plants, one long-time breeder burst out laughing. "Oh!" he said, "there's nothing like that. In none of these other types of breeding are the genes considered foreign. You can make species crosses, generic crosses. Those aren't foreign. Even somaclones—somatic hybrids. Even a hybrid between a mouse cell and a tomato cell is not policed. There's not a legal thing you have to do. Traditionally bred cultivars have no legals."

Nor has anyone ever suggested that mutagenized or irradiated crop seeds—like Maan's experimental hybrid wheat—should be planted behind double fences topped with barbed wire and guarded around the clock. Yet that's how the first genetically modified corn plants were grown at the USDA's Beltsville Agricultural Research Center in the late 1980s. Unlike the surveillance of Parmentier's potatoes, growing outside Paris in 1786, the guard was not meant to convince passersby that the plants were valuable, nor even to protect them from being stolen. It was called for only because the rules for field-testing this new variety of corn were based on outsized fears that crop plants modified using molecular techniques were somehow new and might be dangerous.

Those fears date from the early days of recombinant DNA. By 1972 DNA from almost any source could be cloned in bacterial cells by recombining it with plasmid DNA. Some scientists—among them Paul Berg, who would later receive the Nobel Prize for his pioneering recombinant DNA work—began to worry that, by using these techniques, researchers might unwittingly create a new human pathogen. If it escaped from the laboratory, it could never be recalled.

Berg's lab had developed a way to splice together DNA from different organisms several years before Boyer and Cohen did. But Berg chose not to replicate the molecules in *E. coli* or other bacteria, fearful of the biological consequences. Conversations at meetings of molecular biologists in the summers of 1972 and 1973 often converged intensely on whether recombinant DNA experimentation could unintentionally create hazardous organisms. By 1974 Berg had persuaded Boyer and Cohen, as well as other prominent molecular biologists, to state their concerns in public. They wrote a letter to the National Academy of Sciences and published it in *Science* magazine. It was titled: "Potential biohazards of recombinant DNA molecules."

This self-organized group of concerned scientists formulated four goals: to institute a moratorium on recombinant DNA research, to address public fears about creating new genes, to consider what would happen if such genes got out of the laboratory, and to ask the director of the National Institutes of Health, Robert Stone, to assemble a committee that would write guidelines, carry out risk assessment, and convene an international group of scientific leaders to discuss the guidelines.

Stone quickly formed a 15-member committee, the first Recombinant DNA Advisory Committee, better known by its colorful—and arguably appropriate—acronym as the RAC. An international meeting was convened at the Asilomar Conference Center in California. Scientists from all over the world argued, sought to persuade each other, and eventually crafted a framework for carrying out experiments.

Not everyone agreed that regulations were necessary. Joshua Lederberg, discoverer of bacterial sex, said that just the act of regulating recombinant DNA research would make people think it was dangerous, whether it was or not. Reflecting on what happened in the following decades, Jim Watson was quite blunt, saying: "And boy, he was right."

Donald Fredrickson assumed the directorship of the National Institutes of Health (NIH) in mid-1975 and began to assess his responsibilities. He recalls: "Little did I know what was coming." But from the beginning he was determined to address both the social and the scientific issues. The NIH established an Office of Recombinant DNA Activities and began converting the Asilomar framework into what would become the NIH Guidelines for Recombinant DNA Research. The first version was issued in June 1976. Fredrickson was clear that the guidelines were not regulations. Because they addressed hypothetical, not known, hazards they would evolve as knowledge accumulated and experience grew.

Not unexpectedly, many in the federal government and in communities around the country disagreed. They began to clamor for laws to regulate recombinant DNA research. After all, if scientists themselves had raised questions, surely there was something to worry about. But laws are difficult to undo. The research might prove quite harmless. If it did, laws would impose unnecessary restrictions, interfering with the development of much-needed new pharmaceutical products, such as human insulin, interferon, tissue plasminogen activator, and hepatitis B vaccine. Fredrickson urged President Gerald Ford to extend the NIH guidelines to all federal and private research. After some delay, the directive was issued in September 1976, and the Federal Interagency Committee on Recombinant DNA Research was formed to look at the regulatory authorities of each agency and to see whether new laws were really necessary. Efforts to pass laws did not stop. Over the next several years, 12 bills to regulate DNA research were introduced in Congress; none was passed.

As soon as the NIH guidelines were issued, it was clear that they needed revising. The Asilomar framework, for instance, was limited by the knowledge of those who had attended the meeting. None of these scientists worked with pathogens and the sometimes deadly toxins they produce. So the first guidelines were dominated by the fear that recombinant DNA could turn the laboratory strain of the common gut bacterium *E. coli* into a virulent and contagious pathogen. Experts in pathogens, brought in to educate the Recombinant DNA Advisory Committee, explained that a pathogen was more than simply an organism that can produce a toxin. A pathogen needs a complex delivery mechanism to invade another organism and deliver its toxin to the right cells. Scientists who studied pathogens had learned how to grow them carefully and analyze them in ways that did not jeopardize their own health. They pointed out that recombinant DNA techniques could make their research even safer. Instead of working with the pathogen itself, they could study individual pathogen genes—including toxin genes—by expressing them in a crippled laboratory bacterium that had no capacity to deliver its poison.

Each of the other issues that had aroused early fears was examined in equally careful detail. The number of laboratories and companies carrying out recombinant DNA research expanded rapidly—none of the hypothetical disasters materialized. Confidence in the safety of recombinant DNA steadily grew. Revision by revision, the guidelines were relaxed. More kinds of experiments were classified as exempt, as holding so little potential for harm that they could be carried out under ordinary laboratory conditions. Today we know that, far from being inherently hazardous, recombinant DNA technology is among the safest technologies ever developed.

In a memoir Fredrickson concluded: "It is possible that the 'recombinant DNA affair' will someday be regarded as a social aberration, with the guidelines preserved under glass. Even so, we can say that the beginnings were honorable. Faced with real questions of theoretical risks, the scientists paused and then decided to proceed with caution. That decision gave rise to dangerous overreaction and exploitation, which gravely obstructed the subsequent course. Uncertainty of risk, however, is a compelling reason for caution. It will occur again in some areas of scientific research, and the initial response must be the same."

Yet even as the guidelines for laboratory research were being relaxed, the next area of concern was emerging. An unfortunate turn of phrase—"release into the environment"—had been chosen to describe field tests of plants and animals, including bacteria and other microorganisms, modified by recombinant DNA techniques. Unlike the familiar term "field test," "release experiment" had the unintentional and menacing implication that, once released, these plants or animals could not be controlled.

The NIH approved the first formal release in 1983. As biologist Paul Lurquin writes in *High Tech Harvest: Understanding Genetically Modified Food Plants*, the experiment in which one researcher drank a vial of genetically engineered *E. coli* bacteria "to demonstrate that they could not survive in the human gut" did not count as a release. Nor did the estimated 100 million recombinant organisms that were escaping the world's labs every day on the clothes and shoes of each scientist or technician. Henry Miller, a former FDA official now at Stanford University, noted, "A vast and varied unsupervised 'release experiment' involving tens of thousands of laboratories and untold millions of discrete new genotypes of recombinant microorganisms has been in progress for three decades, with no known untoward results."

The first organism officially released was an altered strain of a common bacterium, *Pseudomonas syringae*. The normal *Pseudomonas* is a nuisance to farmers. It damages crop plants by causing ice to form on their leaves. Just as clouds form by the condensation of water around particles of dust in the sky, ice forms more readily when nucleated by a particle, even a protein. Certain bacteria on leaves produce a particularly efficient ice-nucleating protein. *Pseudomonas syringae* contains a gene, the *ice+* gene, that encodes the ice-nucleating protein. Plant pathologist Steven Lindow of the University of California, Berkeley, isolated the gene, deleted much of it, and reintroduced it into the bacterium. Then he sprayed his "ice-minus" strain on strawberry plants to see if the temperature at which ice formed could be reduced by a crucial few degrees, letting farmers protect their strawberries from late spring frosts.

Because of state regulations the researcher who did the spraying, Julianne Lindemann, wore a "moon suit"—a full-body white coverall, with helmet and breathing apparatus. Her picture became an icon for protesters against recombinant DNA technology, in spite of the fact that it implied danger where none existed. It was taken, as one biologist points out, "by unprotected photographers standing some 10 feet away"—photographers who saw no danger to themselves in being exposed to ice-minus bacteria. (Ice-minus variants of *P. syringae* are, in fact, common in nature, comprising about half of the population.) Afterwards, activists made a practice of tearing out Lindow's strawberry plots, and work on ice-minus was brought to a halt. "The unbelievable irony," writes Lurquin, "is that the bacteria he modified to prevent ice formation in plants are used today in large amounts in their normal configuration (that is, containing the ice nucleation gene) to generate snow on ski slopes during warm winter weather. I wonder sometimes if the people who destroyed Lindow's test plots strapped on their skis right after their acts of vandalism."

The NIH was reprimanded by a federal court for approving the release of ice-minus bacteria without filing a formal environmental assessment, and in early 1984 a Cabinet Council Working Group was formed to bring the federal agencies' representatives together to work out the regulatory issues. It proposed a Coordinated Framework for the Regulation of Biotechnology, published in 1986. The agencies' efforts to avoid new legislation began a process with outcomes that few people—except perhaps Lederberg—foresaw.

The regulation of plants and plant foods fell to three federal agencies. Under the Federal Plant Pest Act of 1957 and the Plant Quarantine Act of 1912, the USDA has the power to decide if a new plant variety or microorganism is likely to become a pest. Its Animal and

Plant Health Inspection Service (APHIS) keeps a list of actual plant pests, on which can be found both *Agrobacterium tumefasciens* and cauliflower mosaic virus. Consequently, all plant varieties created using *Agrobacterium*-mediated transformation or the CaMV 35S promoter—which includes almost all of those that have come to be labelled genetically modified—were defined as "regulated articles." The term was coined by the USDA to capture, for case-by-case regulation, not only bona fide plant pathogens, but organisms that "may be" plant pests.

The USDA further extended that definition to include any organism that contains DNA from different genera. This extended definition covers all plant varieties created through recombinant DNA techniques and *Agrobacterium*-mediated transformation. Plants produced using wide crosses between genera or the techniques Maan used to create hybrid wheat are not, however, regulated. A breeder using molecular modification techniques must obtain a permit for field tests and, after several years of tests, petition APHIS to "deregulate" the new crop, that is, to take it off the list of regulated articles. The petition must describe the genes, regulatory sequences, and transformation procedure used, analyze the plant's genetic and agronomic traits, and provide data on any environmental consequences of growing the plant. APHIS then performs an environmental assessment. If the agency decides that a new variety poses an environmental risk, it can halt all further development.

The second federal agency, the EPA, likewise identified two existing statutes that could apply. These were the Toxic Substance Control Act (TSCA) and the Federal Insecticide, Fungicide, and Rodenticide Act (FIFRA). TSCA was passed in 1976 to allow EPA to track industrial chemicals. FIFRA, passed in 1947, lets the EPA assess the toxicity of chemicals and living organisms used to control fungi, insect, and animal pests and set "tolerances," the amounts of such chemicals that can persist in the environment or in foods without creating a health hazard. To stretch these statutes to cover plants and animals modified by recombinant DNA techniques, the EPA found it necessary to define such organisms as "new." An organism would be "new" if significant human intervention had been used to develop it. (In practice, only those modified by molecular techniques are considered "new," although it would seem that the amount of intervention Maan used to develop hybrid wheat, for instance, is "significant.")

Under these two acts the EPA regulates plants, such as Bt corn or virus-resistant papaya, modified to protect themselves from insects or viruses. It regulates microorganisms, such as the ice-minus bacterium or one, under development, that would allow a plant other than a legume to fix nitrogen. And it is concerned with new uses of existing pesticides, such as the pairing of Roundup (glyphosate) with Roundup Ready soybeans—but only if the herbicide-resistant plant is produced using molecular techniques. Herbicide-resistant crops created by chemical mutation or by the somaclonal variation that results from conventional tissue-culture techniques—and several such crops are now on the market—are not regulated. The EPA examines data provided by the plant breeder on the inserted genes and their products. It reviews the risks and benefits to the environment. It can require the breeder to prepare a resistance management plan, so that insects or diseases do not quickly become resistant to what it used to call a plant pesticide and now defines with the less disquieting name of plant-incorporated protectant or PIP. It checks if the protectant is toxic to animals or humans and, if it is, sets a tolerance level for residues allowed in food.

The mandate of the third agency, the FDA, comes from the Federal Food, Drug, and Cosmetic Act, signed into law in 1938. By the 1940s the FDA had realized that "the vast research efforts needed to assure that all food chemicals were safe was clearly beyond all foreseeable resources," according to a history published by the agency. In the 1950s three amendments to the law "fundamentally changed the character of the U.S. food and drug law: the Pesticide Amendment (1954), the Food Additives Amendment (1958), and the Color Additive Amendments (1960)," the history explains. "With these laws on the books, it could be said for the first time that no substance can legally be introduced into the U.S. food supply unless there has been a prior determination that it is safe. By requiring the manufacturers to do the research, a problem of unmanageable size was made manageable." The onus is on the seller to ensure that the food is safe.

For a crop created through molecular techniques, the FDA recommends that it be compared to a standard variety to determine whether it is substantially equivalent. If the food's nutrient content has been changed—as in Golden Rice, for example, but not the new Hawaiian papaya varieties—the new crop requires a more stringent review, as do any crops that contain an antibiotic resistance marker gene, which is treated as a food additive. (Most plant developers are now using new marker genes that do not rely on antibiotics.) As its guidelines state, "The FDA considers, based on agency scientists' evaluation of the available information, whether any unresolved issues exist regarding the food derived from the new plant variety that would necessitate legal action by the agency if the product were introduced into commerce. Examples of unresolved issues may include, but are not limited to, significantly increased levels of plant toxicants or anti-nutrients, reduction of important nutrients, new allergens, or the presence in the food of an unapproved food additive."

Through legal action the FDA can take a crop or the food derived from it off the market—no matter how it was developed—if it is found to be unsafe. As an FDA publication noted in 2000, "First and foremost, the law simply forbids the marketing of unsafe food. Anyone who violates this provision may be held criminally liable, the food may be seized and destroyed, and the establishment can be required to cease doing business until it complies with the law." For this reason the FDA's guidelines, unlike those of the other two agencies, were not originally mandatory. It was "prudent practice," the agency stated, "for developers of new varieties to consult with the agency on safety and regulatory questions, especially with regard to products developed through new technology."

Through this stretching of definitions familiar crop plants—corn, wheat, rice, cotton—long subjected to a variety of genetic manipulations came under the regulatory purview of the USDA, the EPA, and FDA, but *only* if the genetic technique used to modify them was molecular.

This process-based definition makes no biological sense. As the Council of the National Academy of Sciences pointed out in a 1987 publication, it sets one kind of genetic modification apart from all those that breeders have used for decades. The Council had asked a small group of experts in molecular techniques, ecology, evolution, and plant pathology to examine the various kinds of environmental and health problems that might arise from modifying plants, animals, and microorganisms by recombinant DNA (R-DNA) techniques. The committee reported that the many thousands of plants that had been made using these methods had not revealed unexpected hazards. Indeed, the problems were familiar ones. They were the same as the problems of plants modified by the many other genetic techniques in use. The committee concluded: "Assessment of the risks of introducing R-DNA-engineered organisms into the environment should be based on the nature of the organism and the environment into which it is introduced, not on the method by which it was produced."

But it was already too late. The double standard was firmly in place. Despite wide adoption of this language, and the recognition that regulation should not be process-based, it is.

It took from 1995 to 1997 for the USDA, EPA, and FDA to approve the virus-resistant papaya varieties SunUp and Rainbow even, as Gonsalves writes, "with excellent cooperation from these agencies." Licensing agreements with the several owners of the patented technologies used were reached in April 1998. Papaya seeds were distributed to Hawaiian farmers—free of charge—in May 1998, and the first SunUp and Rainbow papayas were sold in U.S. markets that year. Almost miraculously, the Hawaiian papaya industry bounced back, with production in 2000 nearing levels that hadn't been seen for half a decade. By 2003 75 percent of the papaya grown in Hawaii were Rainbow or SunUp. Under the headline, "Stalked by Deadly Virus, Papaya Lives to Breed Again," the New York Times in July 1999 credited the genetically engineered papayas with having saved the Hawaiian papaya industry.

In spite of this positive publicity, Hawaiian papaya seeds, created by university and government scientists and given away free, are not the genetically modified food that first comes to consumers'—or activists'—minds. As tracked by the International Service for the Acquisition of Agri-biotech Applications (or ISAAA), the genetically modified food grown in 2002 by more than 5 million farmers on 145 million acres in 16 countries worldwide is the product of five large, multinational corporations: Aventis, BASF, Dupont, Monsanto, and Syngenta. Some 90 percent of the market is Monsanto's alone.

The 145 million acres (more than 95 million acres in the U.S.) are planted in varieties of soy, canola, cotton, and corn that can tolerate an herbicide (the list includes the glyphosate in Roundup; the glufosinate in Basta, Challenge, Rely, and Finale; and the sethoxydim in Poast) or that make their own Bt pesticide, a crystalline protein from the bacterium *Bacillus thuringiensis.* Two crops (cotton and corn) can do both. The world leader, with 62 percent of the area farmed, is herbicide-tolerant soy.

So few acres are planted in virus-resistant papaya that they do not register on the ISAAA's charts. Because of their market predominance, Bt corn and Roundup Ready soy, canola, and cotton, all sold by Monsanto, have come to mean GM or GMO in the minds of most protestors against genetically modified foods. Monsanto has been dubbed "Monsatan" by activists in the United Kingdom and has become a scapegoat for the industry. Debating Lord Peter Melchett, an organic farmer and former U.K. Labour minister in 1999, Monsanto C.E.O. Robert B. Shapiro was "surprisingly contrite." He sounded, wrote Michael Specter in the *New Yorker* magazine, "like one of those Chinese leaders who during the Cultural Revolution were made to walk through the streets in a dunce cap. 'Our confidence in this technology and our enthusiasm for it has, I think, widely been seen, and understandably so, as condescension or indeed arrogance,' he said. 'Because we thought it was our job to persuade, too often we forgot to listen.'"

For every other crop technology, from hybrid corn to the latest Intellicoat Early Plant corn—each seed encased in a hot-pink polymer coating that monitors the soil temperature, keeping the seed from sprouting when spring comes late—the company's job has solely been to persuade. Those who needed persuading were farmers, not consumers. Seeds of new GM varieties bought from seed companies cost more than those not modified by molecular techniques. Farmers must agree not to save seeds for replanting the next year. Most farmers do not save seeds of corn, soy, canola, or cotton anyway. The practice hasn't been common in America since the 1930s, because saved seeds often transmit diseases. But the idea of signing a contract to that effect rubs some farmers the wrong way.

Despite the high costs and the contracts, the acreage planted in GM varieties of corn, soy, canola, and cotton continues to grow. From 2001 to 2002 it increased about 10 percent in the U.S., 12 percent worldwide. The reason, according to one analyst, is that farmers found genetically modified seeds "an attractive commercial option. Farmers planting herbicide-tolerant GM soybeans in the United States, for example, could gain roughly $6 per acre in reduced herbicide costs, despite technology fees and no change in yields."

Farmers have not been uniformly satisfied, as with any new technology. Some have run afoul of the intellectual property rules. By saving or selling seed, or even by failing to weed out those seeds that escaped the harvest and sprouted as volunteers the next season, they can bring on Monsanto's private investigators (clued in via the company's toll-free tip line by a law-abiding neighbor). As with teenagers nabbed for pirating music over the Internet, the farmers caught tend to expostulate loudly about freedom. Monsanto, like the rock stars who object to their music being "shared" without royalties being paid, comes off looking like a bully.

Protesters have picked up on these intellectual property disagreements between multinational companies and family farmers to argue that genetic engineering will give control of our seed—our food—to heartless corporations. Yet the protesters too "forgot to listen." Some genetically engineered crops—like the virus-resistant papaya—were designed specifically to solve the problems of small, family farmers. They make no profit for Monsanto, even though Monsanto controls the patent on the coat protein technology; it was donated. The fact that there are so few of these kinds of genetically modified crops is not the result of any failure in the technology. According to many scientists, the reason a handful of companies dominates the market is the cost of complying with federal regulations. Gonsalves and his colleagues persevered; many lack the resources and simply give up.

Ingo Potrykus hopes that the Golden Rice he and Peter Beyer patented will be given away free, and that farmers will be encouraged to save the seeds. He noted in 2003 that activists have "not completely unjustified concerns that big companies might dominate the seed markets and food production. But this has nothing to do with the technology," he said. "If following the regulations were easier, small companies could come up with comparable products. I very much dislike this concentration process," he added, referring to the fact that the vast majority of genetically modified crops being grown were developed by large profit-making companies and not by universities or nonprofit agricultural research institutes. "The opposition is against it, too," Potrykus continued, "but they are the cause of it. They're the ones who have made it so expensive."

To Roger Beachy the regulations are "so impositional that we are really working hard to exclude the public sector, the academic community, from using their skills to improve crops." Noting that the number of field trials run by universities has greatly decreased since 2000, he said, "It's not surprising, because the cost of product development is so high that there is little chance that an academic research team can develop new varieties for commercial release." And it is the universities, the academic researchers, who have historically made a point of breeding local varieties, like the Hawaiian papayas, that are meant to benefit small farms.

"When we brought out the coat protein technology," Beachy said, "APHIS, the health inspection service of the USDA, was interested in it. After a significant review, they said that the new crop varieties represented little or no risk. In fact, at one time they considered that coat protein genes could be deregulated as a class of genes. Vegetables have viruses all the time. We've been eating these viruses and their coat proteins for many, many years." A new papaya that expresses a viral coat protein gene contains, in fact, *less* of the coat protein than an ordinary papaya from a virus-infected grove. And yet the USDA was not the only federal agency involved. Beachy recalled, "The EPA took a different view. They said, A virus is a pest. Therefore, the gene that stops a virus must be considered to be a pesticide. In some ways the EPA has begun to regulate the process *not* the product. The USDA, which was once a stalwart of product-based regulations, is now in danger of becoming more process-related in its policies." Plant breeders, he said, are becoming "seriously discouraged."

The effect of the regulations has been, as Joshua Lederberg foresaw, to make people think the technology is dangerous, whether it truly is or not.

8

IS IT SAFE TO EAT?

Judging the "safety" of food is hardly an exact science. Through experimental testing it is possible to certify that some foods will be dangerous for human consumption, but certifying a complete absence of danger is (like any effort to prove a negative) beyond the capability of experimental science.

—Robert Paarlberg (2001)

Sixty-five percent of the Americans queried for an international survey on genetically modified foods got the answer to the following question wrong: "Do ordinary tomatoes contain genes, or is it only genetically modified tomatoes that do so?" All of our food contains genes—all our plant food and all our animal food. As one biologist explained in a teaching module for middle schools, "A pound of broccoli, for example, contains about a tenth of an ounce of DNA." According to a 1997 study, people eat up to a gram of DNA a day. (One ounce is equivalent to about 28 grams.) Of that, less than 1/250,000 was introduced by genetic engineering. The rest was always in our diet.

When it is eaten, a gene from any source, whether from an animal, an insect, a plant, a virus, or a bacterium, is broken down during digestion into the building blocks of DNA, the nucleotides adenine (A), thymine (T), guanine (G), and cytosine (C). The process begins in the mouth. Saliva contains an enzyme, called deoxyribonuclease, specifically designed to dismember deoxyribonucleic acid, DNA. More of this enzyme is produced in the pancreas and in the small intestine. The

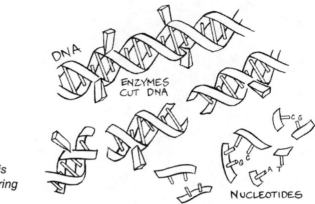

How DNA is degraded during digestion

stomach also contributes to the process: Stomach acid attacks DNA at two of the nucleotides, A and G, inactivating the whole molecule.

The breakdown is not 100 percent efficient. Bits and pieces—short stretches of DNA—escape, a fact that has been trumpeted by certain activists. It might be the source of the Europeans' fears, as documented by another survey, that eating genes will alter their own genes, a concept scientists call "horizontal" gene transfer ("vertical" gene transfer is what happens when you pass your genes on to your children).

Studies showing that bits of genes persist for a while during the digestive process have been done in chickens, sheep, and people, as well as in laboratory mice. The first two digest their food differently from the last two. In chickens, digestion starts in the crop, the stone-filled bag in the bird's throat. Sheep, like people, use saliva at first, but unlike us they have two stomachs.

In chickens, researchers at the University of Leeds in England found an antibiotic-resistance marker gene from genetically modified corn in the crops of all five birds studied. The gene could be found in the stomachs of two of the chickens, but not in the lower intestines of any of them. When the scientists looked for a natural plant gene, they found it in the same places.

The same researchers fed sheep kernels of corn that had been genetically modified to produce the insecticide Bt, a toxin from the bacterium *Bacillus thuringiensis.* Earlier they had sampled fluid from a sheep's first stomach, the rumen, and found it inactivated naked DNA in a test tube within a minute. Five hours after the sheep ate the corn, the scientists could detect in the rumen the gene encoding the Bt toxin. But when the experiment was done using silage only a small fragment of the gene could be found in the rumen, and that no longer than three hours later. Silage is made by letting the corn ferment: by the time the sheep ate it, the plant cells had begun to decay. The DNA was easier to digest because the enzymes could reach it more easily.

A group at the University of Newcastle upon Tyne, led by Harry Gilbert, used human volunteers—though they were not normal, healthy subjects. Each had undergone an ileostomy, in which the ileum, part of the digestive tract, is removed. Each wore a bypass bag, which collects the waste produced by that drastically shortened gut. The scientists fed these seven ileostomists a burger and a shake containing genetically modified soy. After waiting a suitable time, they sampled the contents of the bypass bags, looking for the transgene, which was a gene that made the soy plant resistant to an herbicide. "Whilst the amount of transgene that survived passage from the small bowel was highly variable between subjects, the nucleic acid was detected in all seven subjects," the researchers write. In one person 3.7 percent of the total amount of transgene he or she had eaten made it to the bag. In the others the amount was less. Gilbert's team then repeated their study with healthy volunteers, sampling their feces. "No transgene DNA was detected in the faeces," they write, "indicating that the nucleic acid did not survive passage through the complete intestine." The ileostomy bags had given the researchers a peek into the digestive process. At mid-point in digestion a small percentage of the gene remained whole. By the time digestion was complete the DNA had been broken down.

Walter Doerfler, in the Institute of Genetics at the University of Cologne in Germany, has been studying the fate of DNA eaten by laboratory mice for 10 years. He and his colleagues have reported that some DNA can escape being digested and pass through the intestinal wall. It can enter the bloodstream and interact with the mouse's own cells. The DNA they used at first came from a bacteriophage called M13. It was natural DNA, not a transgene. They detected the M13 DNA in the small intestine, the large intestine, and the blood, as well as in the feces. They found it in spleen and liver cells, and in one out of a thousand white blood cells. They even found it in the cells of a mouse fetus, meaning that it had traveled through the placenta from mother to unborn child.

Doerfler's work has been used as proof that eating genetically modified food is dangerous. Yet, as a report prepared for the U.K.'s Secretary of State for the Environment, Food, and Rural Affairs noted in 2003, "It is important to emphasise that these studies are not focused on transgenes and they are relevant to the fate of all consumed DNA." They apply equally to the tenth of an ounce of DNA in an ordinary pound of broccoli as to the pesticide-coding gene engineered into transgenic Bt corn.

Also important to keep in mind is how Doerfler's methods made it possible to detect even the tiniest bit of foreign DNA. In the M13 experiments the mice were fed 50 micrograms (50 millionths of a gram) of a single, small DNA molecule, one only about 6,400 base pairs long—a very big dose of a very small DNA sequence. In 50 micrograms of M13 DNA the concentration of any one particular segment is much, much higher than in 50 micrograms of, say, the soybean DNA used by Gilbert and his colleagues. The soybean DNA is a billion base pairs long. To sample mouse blood or spleen cells for foreign DNA, Doerfler and his colleagues used the PCR technique, which can amplify the smallest trace of any DNA. PCR doesn't understand the word foreign. You must give it a specific DNA sequence to look for, a list of A's, T's, G's, and C's in exactly the right order. Perhaps that sequence is 1,000 base pairs long. If so, it will represent more than 10 percent of the M13 DNA, but only a millionth of the soybean DNA.

To make it a fair search you'd need to feed a mouse a hundred thousand times more soybean DNA than you fed it M13 DNA. To match 50 micrograms of M13 DNA—an amount that would disappear in the bottom of a tiny salt spoon—you'd have to feed the mouse a couple of teaspoonfuls of pure soybean DNA. Yet people—and mice—do not eat pure soybean DNA. A soybean is mostly starch and protein. To eat a teaspoonful of soybean DNA in the usual way would mean eating pounds and pounds of soybeans—a distinctly unnatural scenario. So feeding a mouse the pure DNA of a simple bacteriophage is not at all like feeding it food from a genetically modified plant.

What these experiments show, however, is that some DNA does escape the process of digestion. A little will incorporate itself into our blood and liver and perhaps even pass through into our unborn children. Does that mean it could alter our genes? Yes, in principle, it could. But has it? People have been eating DNA-rich foods for all of their evolutionary history. Humans have a particularly eclectic diet, consuming foods derived from or containing bacteria and fungi, as well as a great variety of different plants and animals. That is, we've been exposing our digestive tracts to foreign DNA for untold eons before recombinant DNA techniques were invented. Because plant genes, bacterial genes, and animal genes are different from human genes, we should be able to identify the foreign genes in the human genome—if they're there—using computers to compare the vast DNA sequence databases available today.

Animal genes, plant genes, human genes, and even bacterial genes all work by the same rules. How genes are transcribed and translated into proteins is common to all of life. But that's like saying all books contain words. Words belong to different languages. A Japanese word is easy to tell apart from an English one. Even if the two words mean the same thing in both languages, they can be distinguished. Bacterial genes have a different structure from plant genes, and plant genes have a different structure from animal genes. The genes of bacteria are generally not interrupted by introns, for instance, while both plant and animal genes have them. The protein-coding sequences are different as well. A human hemoglobin gene looks very much like a mouse hemoglobin gene, but very different from a plant hemoglobin gene.

Proteins change more slowly than the DNA sequences that encode them, so the correspondence between two genes can often be identified when the protein sequences are compared. So, for instance, when scientists analyzed the DNA in the mitochondria of plant and animal cells and the DNA in the chloroplasts of plant cells, they found that in both structure and sequence these genes were more similar to those of certain bacteria than they were to the genes in the nuclei of animals and plants. This discovery confirmed biologist Lynn Margulis's theory that both mitochondria and chloroplasts were descended from bacteria that had invaded and taken up residence in cells, creating a symbiotic partnership. Other studies found that in plants, at least, the symbiosis is not at all static: genes from the chloroplast, for instance, migrate into the plant's nucleus at a surprisingly high rate.

That we can read in the genes the history of the partnership between chloroplasts and plants means that we should also be able to detect plant, viral, and bacterial genes in the human genome—if they were there. Viral genes are certainly there, but they're not from plant viruses or bacteriophage. They're from viruses that infect animals and humans. It is true that a handful of human genes do look more like bacterial genes than like any other kinds of genes. When the human genome was first sequenced, it was reported that there were about 200 human genes that might have come directly from bacteria—out of the more than 30,000 genes in the human genome. After more investigation that number was reduced by half, then three-quarters. Some of the candidate bacterial genes turned out to be what they appeared to be: genes from bacteria that had contaminated the sequence analysis and were not part of the human genome at all. The experts are still arguing—and probably will for a while—about how to interpret the rest. Are they due to horizontal gene transfer from bacteria to humans? Or do they look out of place because all of their other relatives were lost over the course of evolution? It seems more and more likely that the relatives do exist, but just haven't yet been detected. As more and more animal genomes are sequenced and searched the number of such genes becomes smaller and smaller. Relatives of the candidate genes keep turning up.

So despite all of the opportunities cells have had to exchange genes, genomes maintain their identities. Rather than being surprised that there are bacterial genes in our genomes, we should perhaps be surprised that there are so few of them.

One set of bacterial genes is often singled out in reports on the safety of genetically modified foods—such as the reports by the Society of Toxicologists in 2003, the American Medical Association in 2000, and the World Health Organization and U.N. Food and Agriculture Organization in both 2000 and 2001. These genes are treated as being somehow different, somehow more suspicious. These are genes that make the bacteria resistant to antibiotic drugs.

Antibiotic-resistance genes were, from the beginnings of genetic engineering, commonly used as markers to identify the plant cells that had picked up the new DNA. When Stanley Cohen and Herbert Boyer patented their gene-splicing technique in 1980, they pointed out that a crucial step was being able to separate the transformed cells carrying the recombinant DNA from the parent cells. By linking an antibiotic resistance gene to the gene to be introduced, and then growing the cells in petri dishes containing the antibiotic, it was easy to pick out the cells that got the new genes: they were the only ones that survived on the antibiotic.

Antibiotic resistance wasn't the only marker Cohen and Boyer suggested. Resistance to heavy metals, or an ability to manufacture a certain amino acid or another growth factor, would also mark the transformed cells as different. The antibiotic resistance markers, however, were easy to work with and they became a standard tool. When introducing new genes into plants, the marker gene used most often was one that conferred resistance to kanamycin, an antibiotic rarely used in human medicine.

As crops developed using these new techniques got closer to the marketplace, alarms were raised, both inside the scientific community and by such groups as Consumers Union (publisher of *Consumer Reports* magazine). Could eating antibiotic resistance genes in food increase the rate at which harmful bacteria became resistant to the antibiotics that doctors depend on to cure diseases? The issue wasn't that eating an antibiotic resistance marker gene could make a person resistant to—and therefore unable to be cured by—antibiotics. No one suggested that. The fear was that the ability to survive a dose of antibiotics would be transferred from the person's food to the bacteria that normally live inside a person's gut—again, that hypothetical horizontal gene transfer—and that resistance would then spread from an intestinal bacterium to disease-causing bacteria.

Theoretically, such a possibility exists, conceded a 2000 report on agricultural biotechnology by the subcommittee on basic research of the U.S. House of Representative's Committee on Science. "But it is exceedingly unlikely," the report continued, "because it demands numerous steps, each of which also is highly unlikely." The resistance gene would have to escape the plant cell, yet not be degraded by saliva and stomach acids. It would have to make contact with a bacterium, avoid being cut to pieces by the bacterium's restriction enzymes, and recombine with the bacterial chromosome in just the right place and in just the right way to be inserted.

But that's only the first step, because the intestinal bacterium would then, in a separate horizontal transfer, have to pass that very gene on to a bacterium with a different lifestyle, one that caused a disease that a doctor would need to treat with the drug in question. For the kanamycin-resistance gene, that means tuberculosis, one of the few diseases occasionally treated with kanamycin (which has substantial enough side effects in people to be used only as a drug of last resort). This second horizontal gene transfer, too, is rather unlikely, particularly in countries with modern sanitation systems, where surviving gut bacteria are eliminated in sewage treatment plants.

Considerable effort has been invested in trying to detect such a transfer—so far without success. The U.K.'s GM Science Review Panel, formed by Margaret Beckett, the Secretary of State for the Environment, Food, and Rural Affairs, published a report in 2003 summarizing nearly 20 scientific studies on horizontal gene transfer. In none of them was such a heritable horizontal transfer detected. Yet the perception remains that antibiotic resistance marker genes can move out of food and into a gut bacterium and cause a health problem.

The researchers' response has been quite practical. They have devised different approaches, and identified different markers. For example, the gene from jellyfish that codes for a green fluorescent protein has been used as a marker. Cells that contain the gene coding for the fluorescent protein (together with the gene being introduced) look green under ultraviolet light. Cells that failed to take up and insert the new gene remain dark. Another approach uses the sugar mannose. Ordinary plant cells growing in petri dishes are fed the sugars sucrose or glucose for energy. Transformed cells, by contrast, survive when fed another sugar called mannose, because they have received a gene that can convert mannose into glucose. Without that gene, the cells starve. A third approach is to remove the antibiotic marker after it has served its purpose and before the plant is propagated.

It is worth noting that some of the most common genetically modified food plants never did contain an antibiotic marker. In Roundup Ready and other herbicide-tolerant crops, the herbicide itself was used to select the plant cells that had picked up the new genes and to eliminate the cells that didn't. The herbicide was simply added to the petri dish together with the transformed plant cells. The cells that survived were the ones that had incorporated the herbicide-resistance gene.

Another DNA sequence that has been singled out as being somehow different is the CaMV 35S promoter. The idea that crops produced using the CaMV 35S promoter could cause problems for plants and people seems to have come from an article published by the journal *Microbial Ecology in Health and Disease* in 1999. "Cauliflower Mosaic Viral Promoter—A Recipe for Disaster" was written by Mae-Wan Ho and Angela Ryan of the Department of Biological Sciences in the U.K.'s Open University, together with Joe Cummins, a retired professor of genetics at the University of Western Ontario. It is an opinion piece, which means that it wasn't subjected to peer review, the process by which journals assess the logic and scientific validity of what they publish. Mae-Wan Ho is a founder of the Institute of Science in Society, whose mission, according to its website, is to "work for social responsibility and sustainable approaches in science." Ho's biography on the website says that she has a Ph.D. in biochemistry from Hong Kong University and is "a leading exponent of a new science of the organism which has implications for holistic health and sustainable systems." Among her writings is "The Golden Rice—An Exercise in How Not to Do Science," in which she called Ingo Potrykus's Golden Rice "a useless application." It would not provide better nutrition for the poor, she argued; it was "worse than telling them to eat cake."

The CaMV 35S promoter is widely used. According to Michael Hansen of Consumers Union, "all GE [genetically engineered] crops on the market contain it." The reason is simple: all genes need promoters. Without a promoter, a gene can't be transcribed into messenger RNA. If it is not transcribed, it can't be translated into a protein. That is, a gene simply doesn't work without a promoter. But the transcription machinery in a plant cell is different from that in a bacterial cell; the promoters that work in plant cells are different from those found in bacteria. To add a bacterial gene to a plant requires adding a promoter that will work in a plant cell. The CaMV 35S promoter does. To work well, as Beachy found out in his first experiments with viral coat protein genes, a promoter also has to direct a plant to produce an effective amount of the transcript. The CaMV 35S promoter works well in plants, ensuring that the gene is transcribed often. It is also expressed in many different plant tissues throughout the plant's life. For these reasons it quickly became popular among molecular biologists. It's reliable, so it has been used extensively.

Like Ho, Hansen believes the CaMV 35S promoter is "unstable" and will, in addition to the job it was meant to do, activate some "nasty dormant genetic material." It can cause "large-scale genomic rearrangements," say Ho and her colleagues. It could activate dormant viruses, and perhaps even generate new ones. The "promiscuity" of the CaMV promoter could cause it to trigger cancer. Ho and her colleagues are particularly alarmed by the fact that plants have lots of transposons—jumping genes—in their DNA. They say: "The fact that plants are 'loaded' with potentially mobile elements can only make things worse." Most transposons in plants, as Barbara McClintock found in corn, are silent or asleep—Ho calls them "tamed." She and her colleagues worry that the 35S promoter might somehow awaken them and therefore "destabilize the transgenic DNA," and allow it to "generate more exotic invasive elements." They conclude that all transgenic plants—those in which a gene has been transferred from another organism—should be immediately withdrawn from commercial use. "The available evidence," they say, "clearly indicates that there are serious potential hazards associated with the use of the CaMV promoter." But there is no such evidence.

The finding that frightened Ho and Hansen came from a study published in *Plant Journal* in 1999. The study reported that the CaMV promoter contained a site that is cut-and-pasted with other DNA more often than usual. The researchers called that site a recombination hotspot. The hotspot was very close to the end of the promoter. Breaking the DNA at such a hotspot would most likely disrupt the promoter's activity. But being easily disrupted does not mean that the CaMV promoter is unstable or that it poses a hazard. If the promoter is disrupted—separated from the gene—the gene does not work. The plant does not express the trait, such as virus resistance, sought by the plant breeder. Such a plant, in a breeder's eye, is a loser, a cull. It is discarded.

No matter which promoter is used, precisely where the DNA is inserted into a plant's genome is more or less random. A few insertions affect the plant's ability to grow and be productive. These plants, or their progeny, are discarded as well. For this reason, breeders create many transformed lines. They cross the transformed plants to varieties that perform well, then evaluate the progeny in the greenhouse and the field. They check each generation to make sure that it expresses the new trait. The best plants in the field trials are saved; the rest are discarded. Just as Luther Burbank burned tens of thousands of rejected blackberry bushes, the developer of a virus-resistant squash will grow and select plants for five to ten generations or more, weeding out the under-performers. Among those rejected, obviously, will be any plant in which the inserted DNA causes a lethal mutation. Even if a heterozygote that contains a good copy of the mutated gene grows well, those that have two copies of such a gene will not grow at all or will die before setting seed. Such plant families are discarded. Even small effects on the plant's growth rate or its yield are noticed and, if deleterious, cause the breeder to cull. The new trait must be expressed and it must continue to be expressed in subsequent generations. In fact, one of the biggest problems for breeders is not that crops created through molecular techniques acquire extraordinary new traits—that "nasty dormant genetic material" is turned on—but just the opposite. The newly introduced genes tend to turn off after a few generations.

Could the CaMV 35S promoter activate something nasty? If it is nasty enough to affect the plant's ability to grow and produce a good crop, the plant and its siblings and progeny will be discarded. Could it awaken dormant viruses, as Ho and her colleagues assert? Thousands of copies of dormant viruses have been identified in plant genomes, and their gene sequences have been analyzed. The viruses are defective in ways that make it impossible for a promoter to turn them on. They contain single base changes, and usually short insertions and deletions, that make them gibberish to the protein-making machinery—with or without a strong promoter. Moreover, even if such an "activation"—however unlikely—were to happen in a plant, the result would be that this one plant would suffer from a viral infection. If the virus stunted the plant's growth or reduced its yield, this plant too—or its progeny—would be discarded.

The charge that the CaMV 35S promoter could cause cancer was constructed in the following way. Ho and her colleagues call the CaMV promoter "promiscuous" because it works in plants, yeast, algae, and bacteria. "It has the possibility of promoting inappropriate over-expression of genes in all species to which it happens to be transferred," they claim, adding, "One consequence of such inappropriate over-expression of genes may be cancer." That is quite a leap of logic, and it rests on no facts.

Because the CaMV promoter works in plants, yeast, algae, and bacteria does not mean that it is expressed in human cells. Indeed, it doesn't work particularly well other than in plants. Even if it did work in humans, to cause cancer it would not only have to get into a human cell, it would have to insert itself into the human genome in just the right spot to turn on a cancer gene. Lots of cancer-causing genes have been identified and sequenced, along with their promoters. There are indeed rearrangements of human chromosomes that bring a promoter together with a cancer-causing gene. But none of them are plant promoters, despite the fact that humans have, over the centuries, eaten a lot of plant viruses in their food.

CaMV 35S is naturally found in the cauliflower mosaic virus. A cauliflower—or a cabbage or broccoli or any other of the brassicas—infected with cauliflower mosaic virus often doesn't show any symptoms. The plants might be stunted. Sometimes the veins in the leaves of a cauliflower look clear or are banded in green. On the leaves of certain turnip varieties, particularly one called Just Right, the virus produces clear spots making the leaf look mottled. On cabbages and Chinese cabbages, black specks called pepper spot or fly speck can develop after the vegetable has been picked, although some scientists believe these spots are caused by the turnip mosaic virus, a related virus that often infects the plants at the same time. The virus is spread by aphids from a number of weed hosts, including mustard, penny cress, shepherd's purse, charlock, and chickweed. "Management is difficult," notes a flyer from the University of California at Davis. If the edges of fields are not conscientiously weeded, the crops will most likely be infected. If the cauliflower or broccoli head shows no sign of infection (and generally it doesn't), it is sold, cooked, and eaten, along with its viruses.

Ho and her colleagues assert that the viral DNA, and particularly the CaMV 35S promoter, in a virus-infected cauliflower is somehow different from the CaMV 35S in a genetically modified plant. The CaMV 35S promoter in the latter is "naked," they say, and it is known that human cells take up naked DNA. But that isn't so. The CaMV 35S promoter is much less naked in a plant cell than it is in the original virus. A virus is generally a piece of DNA (or RNA) wrapped in a protein coat. By contrast, the DNA in a plant is hidden inside a cellulose box (the plant cell itself), inside the membrane-bound nucleus, and wrapped up in proteins inside of chromosomes. The CaMV promoter is much better dressed in a plant cell than it was in the virus.

What's more, the CaMV 35S promoter is much more dilute in the plant than it is in the virus. In the virus, the CaMV 35S promoter constitutes roughly 10 percent of the virus's whole genome. (The length of the promoter used varies between 350 and 1,200 base pairs; the whole CaMV genome is 8,000 base pairs.) Put into corn, the same promoter is only about 0.004 percent of the 2.5 billion base pairs in the corn genome. In other words, one cauliflower cell might release hundreds of viruses, and each of those virus particles will contain a CaMV 35S promoter. But if that promoter were in a corn cell instead, that cell would have just two copies: one in the paternal genome and one in the maternal genome. So on both grounds, eating a virus-infected head of cauliflower brings us into contact with much more of the CaMV 35S

promoter than eating an ear of genetically modified corn. Eating cauliflower isn't high on the list of cancer-causing behaviors.

If eating genes isn't hazardous, then, are genetically modified foods
safe to eat? That depends on the gene or genes added. But more importantly, it depends on the food itself. As political scientist Robert
Paarlberg notes in his book *The Politics of Precaution: Genetically Modified Crops in Developing Countries,* "Eating any food can be dangerous."

Many familiar foods naturally contain toxic chemicals—poisons
the plant uses to defend itself against insects and browsing animals.
Lima beans contain a chemical that breaks down during digestion into
hydrogen cyanide, which is poisonous. Toxic psoralens in celery cause
skin rashes. Moreover, psoralen cross-links the strands of DNA to each
other, which can cause cancer. A chemical in cauliflower can make the
thyroid enlarge. Carrots contain a nerve poison and a hallucinogen.
Peaches and pears promote goiters. Strawberries contain a chemical
that prevents blood from clotting and can lead to uncontrollable
bleeding. Peas, beans, cereals, and potatoes contain lectins, which cause
nausea, vomiting, and diarrhea, as do the glycoalkaloids also found in
potatoes. Mushrooms, squash, cucumbers, chickpeas, mustard, manioc,
olives, coffee, and tea all contain chemicals that are toxic to humans.

Yet people generally know from experience how to prepare and eat
these foods safely. Few even know about the toxins, because they are
present at levels too low to harm someone eating a varied diet. (You
would have to eat 400 carrots at a time to receive a toxic dose of nerve
poison.) But some foods contain enough of a harmful chemical to be
toxic in the amounts ordinarily consumed. By trial and error over millennia, people have learned to eliminate the toxins while preparing the
food. In the Andes bitter potatoes are detoxified through freeze-drying. In Mexico and the American Southwest they are dipped in clay.
Both methods keep the glycoalkaloids from causing stomach pains and
vomiting.

Manioc, from which cassava meal is ground, "is more poisonous
than the potato," noted Jack Harlan in *The Living Fields.* "It contains
cyanogenic glycosides which when broken down by enzymes release
prussic acid, HCN, one of the most deadly compounds known to man."

To be safely eaten, a manioc tuber must be peeled, grated, and pressed to expel the juice. The meal must be sun dried, fermented, or heated overnight until all possible HCN is formed. Then, when the cassava cake is cooked, the HCN is destroyed by the heat. To eat olives or olive oil, the extremely bitter oleuropein must first be pressed out—the ancient Romans used it as a weed killer and insecticide; it is now considered a pollutant. Yet olive oil, cassava, and potatoes are not only considered to be safe foods, they are staple foods on which cultures depend.

Because proving "a complete *absence* of danger is," in Paarlberg's words, "beyond the capability of experimental science," the practical definition of a safe food has long been based on experience. "If a food has been a familiar component of the human diet for some time without any known adverse effects," Paarlberg explains, "it comes to be 'generally recognized as safe'—or GRAS, to use the terminology of the United States Food and Drug Administration (FDA)."

A novel food can achieve GRAS status if it can be shown to be chemically equivalent to a food with which we are already familiar. Such was the case with canola oil when it was approved for sale in the United States in 1987. Canola oil was the result of a conventional breeding project to modify rapeseed, *Brassica napus*. Cooking oils pressed from the seeds had been popular until nutritional experiments in the 1940s showed that erucic acid, one of the major fatty acids in rapeseed oil, was toxic. By 1968 Canadian plant breeders had reduced the levels of erucic acid in rapeseed oil dramatically. In 1974, having coined the term canola (from Canadian Oil) to describe varieties containing less than 2 percent erucic acid, they began the process of marketing canola oil in the United States.

Yet the new canola oil was not equivalent to the familiar—and harmful—rapeseed oil. Nor did it compare exactly with any other common vegetable oil, because oils vary considerably depending on the plant and its growing conditions. Instead the Canadian breeders compared the individual fatty acids in canola oil to the fatty acids in soy, corn, peanut, safflower, sunflower, and olive oils. They checked toxicological databases and did feeding studies in both animals and humans. They—and the FDA, which accorded it GRAS status in 1987—concluded that canola oil was safe.

Canola oil was used as a case study when the question of how best to assess the safety of new genetically modified foods was discussed by the international working group called together by the Organization for Economic Cooperation and Development (OECD) in the early 1990s. Thirty countries belong to the OECD, including 23 in Europe, Australia, New Zealand, Japan, Korea, Mexico, Canada, and the United States. The process they agreed upon depends on the idea of familiarity or, to use the FDA's term, substantial equivalence. As members of the Food Directorate of Health Canada recently wrote, "Stated most simply, substantial equivalence encourages investigators to compare a product which they have to assess with one with which they are already familiar."

Substantial equivalence is often misunderstood. For example, Michael Pollan in the popular book *The Botany of Desire* writes, "The Food and Drug Administration told me that, because it operates on the assumption that genetically modified plants are 'substantially equivalent' to ordinary plants, the regulation of these foods has been voluntary since 1992." Pollan's errors are common ones. As a 2002 handbook for scientists, *Genetically Modified Crops: Assessing Safety*, points out, such criticisms follow from "the mistaken perception that the determination of substantial equivalence was the end point of a safety assessment rather than the starting point." The end point in the United States is GRAS status—the food is generally recognized as safe. To reach that point plant breeders compare their new crop with one with which they are familiar, that is, a popular variety of the same plant. They look for differences in how well and where the two crops grow, as well as for changes in levels of nutrients and toxic chemicals in the food. As a group of botanists and nutritionists writing in *Nature Biotechnology* pointed out in 2002, "Only those plants that meet the most stringent performance and safety criteria will advance to the regulators' desks where the results of the safety studies will be independently assessed."

The first part of this assessment is one that plant breeders have been performing—voluntarily—for years. (The fact that the assessment is voluntary has nothing to do with the idea of substantial equivalence, as Pollan implied.) When the colorful sweet corn called Indian Summer—whose red, white, yellow, and purple kernels intensify in color when cooked—and the gourmet orange-fleshed watermelon called New Queen were developed in the late 1990s, for example, they were tested through voluntary field trials overseen by All-America Selections. The judges were horticultural professionals; the site of each trial was a garden at a university, a seed company, or a horticultural institution. There Indian Summer and New Queen were grown side by side with the crops most like them then on the market. The judges compared such traits as yield, time of harvest, and how easy they were to grow. They looked at the plants' resistance to pests and diseases, and marked down idiosyncrasies about each plant's shape and growth habits (for instance, New Queen needs 9 feet of garden space, while Indian Summer has to be isolated from any other corn pollen). They judged each crop's taste by eating it and its quality by how it looks and keeps. A crop that outperforms its closest market rival—as both Indian Summer and New Queen did—is given the stamp of an All-America Selections Winner.

The second step in determining substantial equivalence is new. As the authors of *Genetically Modified Crops: Assessing Safety* note, "Normally, new varieties of foods and crops have not been subjected to traditional toxicological testing." Determining if the bright colors of Indian Summer and New Queen marked a rise—or fall—in the level of a nutrient or a toxic chemical was not part of the All-America Selections trial or any other official test. As Jim Maryanski of the FDA notes, "This is in contrast to what is done with engineered varieties, where there are far more tests being done for nutrients, toxins, vitamins and minerals, and so forth."

In the late 1960s Wilford Mills of the Pennsylvania State University crossed the popular potato variety Delta Gold with a wild potato relative from Peru. He called the new variety Lenape. The wild genes made Lenape highly resistant to attack by insects and potato blight. "Lenape was a wonderful potato," remembers Mills's colleague Herb Cole. "It chipped golden." A Pennsylvania potato chip manufacturer earmarked it as a favorite. A potato breeder in Ontario thought Lenape would be a good variety for early potatoes, the kind harvested young and boiled with peas. "He cooked up a batch of potatoes and peas," Cole said, telling the story in 2003, "and he got very nauseous. He figured it was just an accident, so he cooked up some more the next week. He got even sicker." He asked a biochemist at his university to analyze the potatoes. It turned out that they were exceedingly high in glycoalkaloids, the potato's natural toxins. He reported the results to Mills, who enlisted Cole in analyzing the variety further. "The result was that Penn State recalled the Lenape potato," Cole said.

Lenape had been released as a public variety, not patented. Recalling it meant contacting every grower of seed potatoes and requesting that they not market Lenape seed tubers. Cole and Mills did so. "But some of Lenape's heritage has carried forward and been bred into other varieties used today," Cole added. A list of potato cultivars compiled by the crop and soil science department of Michigan State University includes Lenape in the parentage of 13 varieties. "In making the cross," Cole concluded, "Bill did what all the people opposed to biotechnology say you ought to do. He went back to the origins of the potato and brought along genes for insect and disease resistance. He also brought along genes for glycoalkaloids." But he didn't know it.

There were no such surprises when Roundup Ready soybeans were first eaten. Before any were marketed they were compared with regular soybeans to see if there were differences either in the raw seeds or in toasted soybean meal. Monsanto's researchers checked the fatty acid composition of the oil and its total quantity. They looked at the amounts of fiber, ash, and water, and compared the carbohydrates and proteins. They gave special attention to chemicals typically found in soybeans that could be toxic at higher levels or act as antinutrients. They did feeding studies in rats, chickens, catfish, and cattle to show that there were no nutritional differences between Roundup Ready soy and its market rivals. They found no substantial differences.

Other scientists reported in 1999 that Roundup Ready soybeans contained 12 to 14 percent less of a substance, called isoflavone, that might help prevent heart disease, breast cancer, and osteoporosis. Still a third group of scientists, though, found that Roundup Ready soy contained too much isoflavone—enough to cause uterine cancer in laboratory mice. According to scientists from Colorado State University, the difference among the three results is due to environmental conditions, "the kind of variation in isoflavone levels that can occur from year to year. . . . Wine afficionados know that the weather can influence the quality of grapes, causing 'good' years and 'bad' years for wine." The same is true for soybeans. Monsanto's experiments are the most trustworthy, in terms of assessing the safety of the crop, the Colorado State researchers confirmed, because Monsanto grew both kinds of soybeans, the conventional ones and the genetically engineered ones, in the same fields. With the two types growing side by side, the effects of the weather can be discounted. Neither of the other experiments compared crops from the same field grown in the same season. As *Genetically Modified Crops: Assessing Safety* concludes, "The choice of the comparator is therefore crucial to the effective application of substantial equivalence in establishing the safety of a GMO-derived food."

Some critics of genetically modified food, such as Michael Hansen of Consumers Union, want to see plant breeders do more. In a chapter in the 2001 volume *Genetically Modified Organisms in Agriculture: Economics and Politics,* he argued that the FDA should require plant breeders to identify the total number of copies of each gene they insert into a plant, the location of each one on a chromosome, the structure of each insert, including a complete genetic map and the full DNA sequence, and the sequence of at least 10,000 base pairs of DNA on either side of it. He further believes that the FDA must demand evidence of both the structural and functional stability of the insert over multiple plant generations. In 2003 the FDA included Hansen on a committee to review their guidelines. Jim Maryanski was asked specifically if there was any case, out of the 50 or more that the FDA has dealt with so far, in which knowledge about the insertion site said anything of value about food safety. Maryanski's answer was a resounding "No."

To someone familiar with the genomes of plants, suggesting such requirements is puzzling. Even two inbred strains of the same crop will show differences in their gene sequences. Rearrangements and transpositions are quite common. As for stability, hybrids—whether created through new methods or old—are not stable over time. After a while they decline in performance—leaving the market looking for the next new variety. Wide crosses, too, trigger instability. Nor are these necessarily the work of scientists and plant breeders. Some simply happen naturally, such as the distant crosses that gave rise to our contemporary strains of wheat. Avraham Levy and his collaborators at the Weizmann Institute in Israel have recreated those crosses and studied what happens to the genes and the genomes. They found many genetic changes in these wide hybrids: genes shut down, transposons were activated, and many genes were simply deleted.

At the molecular level, instability is a fact of life. Making sure the DNA around the new gene doesn't change—as demanded by Hansen—doesn't provide any information about the safety of the food. The FDA advisory committee came to the conclusion that having more information about the inserted gene and the DNA sequence surrounding it wasn't nearly as important as having more information about the chemical composition of the food produced from the plant.

Indeed, plant geneticists have been surprised at how few gene insertions have a visible effect on plants. Scientists studying the little laboratory weed *Arabidopsis thaliana* have used *Agrobacterium* to make thousands upon thousands of T-DNA insertions in and near genes to see what effect disrupting a particular sequence has on the plant. Most of the time, inserting a new gene has no effect at all; the plants grow normally. A few insertions cause mutations, but for every hundred only one or two make a visible change in how the plant looks, grows, or reproduces.

Gene-activation libraries have been made in which the T-DNA has a CaMV 35S promoter at its very end to activate genes. In these experiments scientists are doing quite deliberately what Mae-Wan Ho feared the 35S promoter might do on its own. To the investigators' disappointment, most of the insertions don't have any effect. The same is true of transposon insertions—most don't cause much change. The vast majority of plants with new insertions—and their offspring, too—are quite normal.

How a plant grows and reproduces is simply not as sensitive to the disruption or activation of any one gene as we might fear. These experiments tell us—if we haven't yet learned it from irradiating plants, culturing them, and treating them with chemical mutagens—that plants and their information storage systems are quite robust. A gene's location, while not unimportant, is less important than what the gene is and what it codes for. Nor is the loss of a gene necessarily a catastrophe. Many genes are represented by several copies—sometimes hundreds of copies. Though each copy might be a bit different, they can often cover for each another so that the loss of one copy goes unnoticed. There can also be more than one way to carry out a needed biochemical function; if one route is blocked, another is often available.

The contemporary molecular plant breeder can be confident that a corn plant with an extra gene is still a corn plant—a fact that Luther Burbank simply took for granted. And if an insertion is bad for the plant, the contemporary breeder does precisely what Burbank did—throws the plant out. But modern breeders have analytical tools available to them that Burbank didn't have. They can analyze the proteins, nucleic acids, fats, and starches, as well as all of the many molecules (called secondary metabolites) that the genetically modified plant makes. They can compare this analysis to the chemical profile of the variety from which the plant was derived and ask, Is this the same plant, except for that one protein encoded by the added gene?

Calgene's FlavrSavr tomato was the first genetically modified whole food. When Calgene brought it to the FDA in 1992, the tomato was subjected to $2 million-worth of testing by the FDA on top of the testing done by Calgene. In a public meeting the FDA scientists brought the results of their extensive and sophisticated chemical analyses to a panel of external advisors; the panel included representatives of public interest groups and industry, as well as scientists whose specialties ranged from nutrition to basic plant science. The concluding slide of the FDA's presentation had a simple message: Calgene's transgenic tomato . . . is a tomato.

9

POISONED RATS OR POISONED WELLS

The increased public awareness of food allergy has arisen from a combination of three factors: reasoned concern, fear through ignorance, and political motivation.

—Bob Buchanan (2001)

On August 10, 1998, Arpad Pusztai was interviewed on the British TV show "World in Action." Pusztai studied lectins, sugar-binding proteins found in peas and beans, cereals and potatoes. In his 35 years at the Rowett Research Institute in Aberdeen, Scotland, he had written three books on lectins and published 270 research papers. Because of his expertise, he had been asked to test the safety of a potato variety that had been genetically engineered to produce its own pesticide. That pesticide was a lectin from a flower, the snowdrop.

Pusztai fed the genetically modified potatoes to rats. His experiments, he told the television audience, showed that the GM potatoes damaged the rats' immune systems and stunted their growth. He himself would not eat GM food, Pusztai said. He found it "very, very unfair to use our fellow citizens as guinea pigs."

Pusztai's study made headlines around the world. It came to be called "the poisoned rat debate." According to a news report in the journal *Science*, the Rowett Institute was flooded with calls from reporters even before the television show aired. The Institute was faced with "a megacrisis we didn't remotely anticipate," said director Philip James. When James examined Pusztai's experiments, he found them a "muddle." Pusztai's laboratory was sealed, his notebooks were turned over to an audit committee, and Pusztai himself was put on indefinite leave—he was out of a job. The audit committee's report, released in October 1998, concluded that Pusztai's data did not support his conclusions.

Forbidden to talk to the press, Pusztai asked a number of scientists to review the audit committee's report. In February 1999 they posted a memorandum, supported by more than 20 scientists, on the World Socialist Web Site. "Those of us who have known Dr. Pusztai's work or have collaborated with him," the memorandum stated, "were shocked by the harshness of his treatment by the Rowett and even more by the impenetrable secrecy surrounding these events. It is an unacceptable code of practice by the Rowett and its director, Professor James, to set themselves up as arbiters or judges of the validity of the data which could have such a profound importance not only for scientists, but also for the public and its health."

One scientist stressed that "there is enough strong evidence that the work of the audit group was not objective and per se dangerous, not only for Dr. Pusztai, but generally for free and objective science." Another thought the Rowett Institute's treatment of Pusztai was "a great injustice." Even Kenneth Lough, who had been the Rowett's principal scientific officer from 1956 to 1987, said, "The Institute is at risk in sending the wrong signals to scientists in this field of research that any sign of apparent default [that is, any error in judgment] will be treated with the utmost severity. The awareness will of course act as a strong deterrent to those who wish to conduct research in this vitally important field."

Pusztai's study was also reviewed by a committee of six members of the British Royal Society. They sent out material from Pusztai, the Rowett Institute, and other sources to scientists with expertise in statistics, clinical trials, physiology, nutrition, quantitative genetics, growth and development, and immunology. The committee concluded that Pusztai's experiments were poorly designed; the statistics he used were inappropriate and his results were inconsistent. They recommended that the experiments be repeated and the results be published.

Pusztai jumped to his own defense with a detailed response circulated on the Internet. He and a colleague with whom he had worked for several years published their original study in the medical journal *Lancet*. For this *Lancet* also came under criticism. The U.K.'s Biotechnology and Biological Sciences Research Council called the journal "irresponsible." *Lancet*'s editor, Richard Horton, stood by his decision. Five of six reviewers had favored publication, and he believed that it was appropriate for the information to be available in the public domain.

The lectin in question is called the *Galanthus nivalis* agglutinin, after the Latin name of the snowdrop, *Galanthus nivalis*. It is abbreviated GNA. Like other lectins, it recognizes and binds to sugars on other proteins. Although lectins were first discovered in plants, we now know that there are many different kinds of lectins in animals as well. Many proteins—in all kinds of organisms—are decorated with sugar molecules. Some, called glycoproteins, carry long strings or branches made of several sugar molecules. Each glycoprotein has a different complement of sugar molecules, depending on what it does and where. The sugar signature works like a zip code in the cell, determining where the protein is delivered by the machinery that produces it.

Lectins read these sugar codes. They serve many functions, one of which is to recognize disease-causing bacteria and viruses. For plants, lectins are a defense against insects. GNA, for instance, is mildly toxic to some pests of rice and other important crops. It does not affect ladybird beetles, considered to be beneficial insects, although it does affect parasitic wasps, which are also beneficial. Other lectins, including one called ricin, are quite toxic. When taken up by cells, ricin blocks the synthesis of proteins by inactivating the ribosomes. GNA does not have this property. Pusztai's own studies showed that rats could safely eat purified GNA. Moreover, he and his colleagues found that GNA protected the rats against infection by *Salmonella* bacteria, the intestinal bug often found in raw eggs and on uncooked chicken. When the gene coding for GNA was introduced into potatoes and rice, it increased the plants' resistance to insect pests. The next question was how a plant that produced GNA would affect a human gut. This question Puzstai tried to answer using rats as stand-ins for humans.

People already eat lectins. They are present in most plants, and are especially abundant in seeds, including cereals and beans, and in tubers such as potatoes. They tend to survive cooking and digestive enzymes. They occasionally cause symptoms of food poisoning. A lectin called phytohemagglutinin or PHA, for example, is a normal component of kidney beans. Allergist David Freed recounts an incident in 1988 when a hospital had a "healthy eating day" in its cafeteria at lunchtime. Thirty-one portions of a dish containing kidney beans were served. Over the next several hours, 11 people experienced typical food-poisoning symptoms, including vomiting and diarrhea. All recovered by the next day, but no pathogen was found in the food. It turned out that the beans contained an abnormally high concentration of PHA.

Studying PHA in the early 1990s, Pusztai and his colleagues found that it caused the cells lining the surfaces of a rat's intestines to die off and be replaced more quickly than usual. The rats became more susceptible to an overgrowth of the common gut bacterium, *Escherichia coli*. *E. coli* is harmless in small numbers, but causes stomach upset if it multiplies. The younger replacement cells on the tiny surface projections, or villi, of the intestines, Pusztai and his colleagues found, had a high proportion of proteins with mannose sugars on the ends of their sugar signatures. Because *E. coli* has projections, or fimbrae, that recognize and bind to mannose, the bacterium could grow more easily. (Pusztai also found that including the lectin GNA in the rat's diet reduced the extent of bacterial overgrowth, because the GNA binds to the mannose and keeps the bacteria's fimbrae from using it.)

As they do in insects, some lectins can get into and through animal cells and enter the bloodstream. Some are potent allergens. So even though GNA appeared to be relatively benign, there was no doubt that a food containing it needed careful testing. Sensibly, the Scottish Office of the Agriculture, Environment, and Fisheries Department commissioned a three-year study in 1995. The University of Durham and the Scottish Crop Research Institute were to provide the transformed potatoes, and the Rowett Institute was to do a chemical analysis of them. Pusztai was also to do both short-term (10-day) and long-term (3-month) rat feeding trials. In these he would compare the effects of eating potatoes from the transgenic plants with the effects of eating potatoes from the parent lines, that is, from the plants which provided the cells transformed by recombinant DNA.

The rats in Pusztai's study were fed either raw or cooked potatoes. If they ate ordinary, non-transgenic potatoes, their diets were supplemented with pure GNA. The results of the experiment showed that the organs of the rats fed transgenic potatoes weighed significantly less than the same organs from control animals fed non-transgenic potatoes. In this control group—the rats fed purified GNA and ordinary potatoes—the lymphocytes (cells in their immune systems) were more responsive to stimulation by other lectins. By contrast, the lymphocytes were depressed in the animals fed the transgenic potatoes. It seemed that the potatoes expressing the GNA protein were somehow poisonous, while the ordinary potatoes—although spiced with pure GNA—were not.

Since he knew GNA wasn't toxic, Pusztai jumped to the conclusion that the new gene itself—or perhaps the other DNA introduced along with it—was causing the problem. He went public with his conclusion on "World in Action." What his experiments actually showed was that the genetically modified potatoes were different from each other, as well as from their parent lines. When the potatoes were chemically analyzed, the researchers measured total protein concentration, as well as the content of several relevant proteins, including GNA and potato lectin. All of these differed between the unmodified and modified potatoes, as well as between the different lines of modified potatoes. A later study on genetically modified potatoes found the same thing. Rather than the effects of the introduced DNA, what Pusztai was most likely seeing were the effects of somaclonal variation, the variation that arises as a result of tissue culture.

Plant breeders have known about somaclonal variation for decades. The techniques of tissue culture come from the 1950s, when Miller and Skoog identified the ingredient in old herring sperm that would make a plant cell divide and form a callus. Callus culture is commonly an intermediate stage between *Agrobacterium*-mediated transformation and the regeneration of transformed plants. With the right mix of hormones, callus cells divide for a long time; weaned off hormones, the callus turns into a plantlet, with roots, a shoot, and leaves. By 1981 the fact that different regenerated plantlets from a single callus were not identical was well-enough known to be given a name.

Somaclonal variation was both a nuisance and an aid. Through screening, breeders could pick and choose among the changes, discarding the bad ones.

Some of the changes, scientists have since learned, are epigenetic, meaning that the modifications affect the expression of genes, but not their structure. These changes can be caused by differences in how the DNA is methylated. Attaching methyl groups to genes shuts them down, silences them, makes them sleep. Taking the methyl groups off awakens them. These changes, though, tend to right themselves quickly once the plants are propagated sexually or through cuttings. More-stable genetic changes—deletions, insertions, single base changes, and rearrangements—also arise. Therefore, every new plant derived using tissue culture techniques must be evaluated both for how it grows and for its food properties. Potato breeders know to be especially careful of the toxic glycoalkaloids that potatoes naturally produce. These chemicals, which contribute to inflammatory bowel disease, are concentrated—not destroyed—when potatoes are fried.

Pusztai's conclusion that the variation he observed was due to genetic engineering was unwarranted. His mistake proved costly—and not just to his reputation. Pusztai's experiments have been attacked for their small sample sizes, for the use of inappropriate statistics, and for the fact that a diet of raw—or even cooked—potatoes is bad for rats, even if supplemented with a bit of extra protein. But oddly enough in the entire poisoned rat debate no one seems to have seen the central flaw in Pusztai's experiments: the absence of appropriate controls. The control in an experiment is the material that allows a good comparison to be made in order to understand the consequences of the experimental treatment being studied. In Pusztai's experiments the control potatoes had a different breeding history than the transgenic potatoes, so they couldn't be compared directly to those being tested. Only the genetically modified potatoes had undergone tissue culture. To blame the new DNA for the potatoes' effects on his rats, Pusztai needed control plants that had also come out of tissue culture and were exactly the same as the ones being tested—except for the new genes. Pusztai didn't use such plants.

To a fellow nutritionist, Pusztai's conclusions might have seemed justified. A plant breeder, familiar with the tissue culture technique, would have seen instead the signs of somaclonal variation. The tissue culture-derived potatoes were very different from those the experiment had been started with. It is quite likely that it was the changes that occurred in culture that were responsible for Pusztai's results, not the introduced gene. Perhaps we will never know. But it is quite clear that the expertise battle that sprang up around Pusztai's experiments obscured an important point. When plants are engineered to express new proteins that could affect human health—and lectins are clearly in this category—the foodstuffs produced from them must be analyzed carefully.

Plant breeders, too, can have their blind spots. The Lenape potato was on the market—and potato chips were being made out of it—when the toxicity was discovered.

Steven Taylor, head of the University of Nebraska's Department of Food Science and Technology, attended a conference at which Jim Cook, a plant pathologist at Washington State University, described a new wheat variety. In 1988 Cook had overseen the first field test of a genetically modified organism in the Pacific Northwest, a soil microorganism modified to protect wheat from root disease. In the early 1990s he had chaired the 70-member international working group that produced the report "Safety Considerations for Biotechnology Scale-up of Crop Plants" for the OECD. The new wheat he described at the conference was resistant to rust, a devastating fungal disease, the scourge that Norman Borlaug had studied half a century earlier. The wheat had been modified to produce a chitinase, an enzyme that can attack and break open molecules in the cell walls of invading fungi. Taylor asked if Cook knew that chitinases were major allergens. Cook replied that he did not. "I relay that story," Taylor writes, "to show how important it is for the people selecting these genes to become more aware of what's going to happen in the commercialization phase and to, perhaps, address some of those questions early in the process."

In the past, few plant breeders had much awareness of the fine points of food safety or nutrition. The two disciplines are in different university departments. There was little overlap until the risks and benefits of genetically modified food began bringing both to the same conferences. John Beard, a nutritionist at Pennsylvania State University, for example, tells the same sort of story about Ingo Potrykus. "I met Potrykus at a conference," Beard said. "I asked him about the beta carotene in Golden Rice, 'Is it bioavailable?' He got this blank look on his face. 'Is it absorbable?' I asked. He said, 'Why wouldn't it be?'" Beard sent Peter Beyer, Potrykus's colleague, references on cell culture systems with which nutritionists test a nutrient's bioavailability, that is, how easily the body can absorb and use it.

This disconnect between the two fields led some nutritionists to sound the alarm against genetically modified food. Marion Nestle, a nutritionist at New York University, for instance, wrote in the *New England Journal of Medicine* in 1996, "Biotechnology companies might be introducing allergenic proteins from donor organisms into the food supply." In an article in the *Cambridge World History of Food*, published in 2002, she elaborates: "Most biotechnology companies are using microorganisms rather than food plants as gene donors. Although these microbial proteins do not appear to share sequence similarities with known food allergens, few of them have as yet entered the food supply. At the present time, their allergenic potential is uncertain, unpredictable, and untestable."

A plant breeder might quibble that we have, indeed, eaten most of these microbial proteins and genes. People are adventurous eaters. We eat plants and animals, fish, crustaceans, worms, insects, fungi, yeast (the Marmite so beloved in England is a yeast extract), and bacteria of all kinds. Some bacteria we eat accidentally—on unwashed lettuce or spotted apples, for instance. Other kinds give us our wonderful array of yogurts and cheeses. In both cases the bacteria we eat are alive, their genes busily producing microbial proteins. Yet even if the genes biotechnologists use come from yeasts and bacteria that we ordinarily eat, the need to better understand food allergies remains.

Many more people think they have a food allergy than actually do. While a quarter of all Americans will say they are allergic to some food, true allergies are actually relatively uncommon. Only about 2 percent of adults and 5 to 8 percent of children exhibit allergic responses when given skin tests with a range of proteins. Nonetheless, allergies are a serious health concern. Allergic reactions range from annoying to life-threatening.

An allergic response, whether to a bee sting, pollen, or peanuts, commonly involves immunoglobulin E, a key molecule in the immune system. Immunoglobulin E, or IgE, comes in various shapes, each of which is made to match a particular protein. IgE molecules sit on the surfaces of special cells called mast cells and basophils. Mast cells lie just beneath the surface of body tissues that are exposed to the outside world: skin, gut, nasal passages, lungs, and urinary tract. Basophils are in the blood. When the shape of the IgE molecules matches that of a protein they're exposed to, the IgE binds to it. In fact, two IgE molecules have to bind to two parts of the protein, called the allergen, thus linking the two IgE molecules together. This cross-linking triggers the mast cells to release packets of highly irritating substances—histamines and leukotrienes—to incite the body to eject the invader in one way or another. A runny nose, itchy eyes, swollen tongue, tight throat, asthma, hives, vomiting, diarrhea, and other discomforts can result.

People with allergies can produce as much as 10 times the average amount of IgE. This makes their mast cells rather jumpy, releasing floods of inflammatory molecules in response to proteins that other

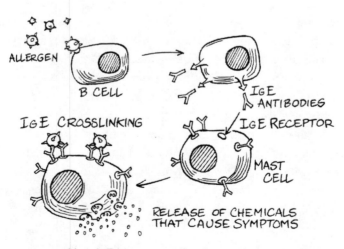

How IgE causes an allergic reaction

people's mast cells ignore. These include proteins in pollen and bee venom as well as in food. And yet just a handful of the many proteins we eat cause most of the problems. "While virtually any food may cause an allergic reaction in someone, somewhere," notes Dean D. Metcalfe, chief of the laboratory of allergic diseases at the NIH, "a relatively small number of foods appear to cause over 90 percent of all reported immediate reactions in both children and adults." In young children, most problems are caused by milk and soybeans. Adults react most to peanuts and other nuts, shrimp and crustaceans, fish, and eggs. Sesame seeds and wheat are sometimes added to this most-allergenic list, as are peaches, plums, apricots, cherries, almonds, celery, and rice. Each of these foods holds hundreds of thousands of proteins. Just one or two of them are usually the IgE triggers, or allergens.

An IgE overreaction can be quite dangerous. Although most people have mild symptoms—hives, diarrhea—allergies can cause anaphylactic shock: difficulty breathing, a rapid and dramatic fall in blood pressure, and cardiovascular complications that can lead to death if not treated immediately. Allergic reactions account for thousands of emergency room visits a year in the U. S. About 135 people die each year as a result of allergic reactions. Although food allergies rarely trigger such extreme responses—the most common cause is bee stings—they can.

For someone who is allergic to a food, avoiding it is the only sure way to be safe. For many American children, that means no peanut butter sandwiches. But avoidance isn't always so easy. Foodstuffs can show up in surprising places. Peanuts lurk in piecrusts, gravy, and hot chocolate, according to a front-page story in the *New York Times* about a promising drug for desensitizing people to peanuts. "One person died after eating an egg roll that used peanut butter to hold the roll together," the *Times* reported.

The FDA recognized the problem of allergenicity early on. In its first guidelines on genetically modified foods, published in 1992, it warned against taking proteins from known allergens, such as peanuts, and moving them into other plants, such as corn. Someone who was allergic to peanuts would never think to avoid cornflakes. A related concern is that a protein from a bacterium or insect, for instance, when moved into a food like corn, might prove to be an allergen.

A 2001 brochure from the radical environmental group Greenpeace echoes Marion Nestle's complaint: "Because most genes being introduced into GE [genetically engineered] plants come from sources which have never been part of the human diet, such as bacteria, insects, and viruses, there is no way of knowing whether or not the products of these genes will cause allergic reactions." John Haglin, a physicist who ran for president of the United States in 2000, raises a new alarm in a video documentary called *Genetically Engineered Food: Are We at Risk?* "If you can teach a tomato, for example, to produce enough flounder blood, that tomato becomes frost-resistant and is more profitable," he said. "What a fabulous application of technology— unless you're allergic to fish, or your children are allergic to fish."

Such proclamations, coupled with the very real concerns of parents whose children have food allergies, have led many people who are not otherwise against innovations to wonder about the wisdom of genetically modified food. And yet conventional plant breeding and many other farming practices can also affect how much of an allergen a crop contains. All allergens are proteins. Both the amount of a particular protein a plant contains and where in the plant that protein is concentrated are influenced by the weather, by stresses due to disease or pests, and by the plant's breeding history. None of these factors has had much effect on food allergies. Instead, says a 2000 report issued by the World Health Organization (WHO) and the U.N.'s Food and Agriculture Organization (FAO), what has caused new allergies to appear are foods eaten in unusual quantities and foods newly brought in from foreign lands. Peanut allergies are frequent in the United States, Canada, and parts of Europe, but not in countries where peanuts are rarely eaten, the report points out. "Also, recent food introductions such as kiwi fruit have proven to be additional sources of food allergens."

In 1962, when the first 2,400 pounds of Chinese gooseberries were sold in Los Angeles under the name kiwi fruit, no one raised the issue of allergenicity. Catherine Woteki is the dean of agriculture at Iowa State University and a former undersecretary of food safety in the USDA. "I can remember the time when there was no kiwi fruit in the U.S.," she said. When the first shipment was brought in, she recalled, "There were no requirements for allergy testing. None." Kiwis were considered "a traditional food," one with a history of having been eaten in other countries. They were assumed to be safe. That designation has not been changed even though many Americans have found they are allergic to kiwi fruit.

Nor is the allergy issue raised today when rambutans and white apricots are turned up by the self-styled Fruit Detective, David Karp, whose job, according to a profile in the *New Yorker* magazine is to "range around the country and the world and find exotic fruits, or uncommon varieties of common fruits" for grocers who like to "titil-late" their customers with "curiosities." The white apricots were cre-ated by a California farmer who had worked for 30 years to interbreed apricot stock he had imported from Morocco and Iran. The rambu-tans, "bright-red, golf-ball-size, tendril-covered fruits from Southeast Asia, with translucent, sweet-tart flesh," were grown in Hawaii. They could not be imported to the mainland until a new technology involv-ing electron beams was devised to destroy the tropical pests they har-bor. "The new technology, many in the exotic-fruit world believe, will greatly expand Americans' awareness of the fruit that the rest of the world eats, and bring a cornucopia of new items to the produce de-partment," the *New Yorker* reported. None of these new introductions will be tested for allergenicity.

The number one food allergen, the peanut, is itself a rather recent newcomer to North America. Although the peanut originated in South America, it came to the United States by way of Africa, having been brought to Virginia by slaves. Its popularity didn't climb until the 1890s, when John Harvey Kellogg, the inventor of cornflakes and granola, began making peanut butter at his health sanitarium. He wanted his patients with poor teeth to enjoy what he called "the noble nut." The fact that peanut butter can cause fatal allergic reactions was then—and is now—generally overlooked. Peanut butter is more likely to be thought of today as a health food, a staple of children's lunchboxes.

But the FDA and developers of genetically modified crops are not ignoring the problem of allergenicity. They consider it their most im-portant food safety issue. And while Greenpeace worries that "there is no way of knowing whether or not the products of these genes will cause allergic reactions," scientists have made considerable progress in identifying and categorizing what makes proteins allergenic.

Most allergenic proteins have several traits in common. They are generally abundant in the food, representing more than 1 percent of the total protein. They survive cooking. And they are quite stable, resisting being broken down by stomach acid and digestive enzymes. But these are not hard and fast rules and there are exceptions. Some very abundant proteins aren't allergens. Some proteins are allergenic, but not very abundant. Some unstable proteins—and even unfolded proteins, which should digest very easily—are allergenic. Yet despite our inability to develop criteria with no exceptions, there is a rapidly growing body of information about what makes a protein an allergen. This information includes the amino acid sequences that interact with IgE, how the protein folds, and the extent to which a protein is decorated with sugar molecules.

To see if a protein is likely to be allergenic, for example, its amino acid sequence is compared to those of known allergens in chunks as small as eight amino acids long. The known allergens with which it is compared include not only food allergens, but proteins from pollens, fungal spores, insect venom, and other things people have allergies to. If a sequence of 80 amino acids or more is the same over a third of this stretch, as a sequence in a known allergen, the protein is considered suspect.

The function of the protein is also a factor. As the WHO-FAO report said in 2000, "Certain classes of protein are well known allergens. For example, the 2S high-methionine albumins from Brazil nut, walnut, sunflower seed, and mustard are major allergens from those sources. Thus, other 2S high-methionine albumins should be scrutinized very carefully." The report suggested that a list be compiled of proteins that scientists agree have "allergenic potential." Such lists have since been made available for everyone's use on the Internet.

If the protein comes from a known allergen, such as the peanut, it can be tested against serum from people who are allergic to peanuts. A test tube assay will show if, and how strongly, each person's IgE reacts to the new protein. If the laboratory test is negative, skin-prick tests—the kind of test done by a doctor to diagnose an allergy in the first place—can be done. If those are negative, allergic people can be recruited to perform a double-blind, placebo-controlled food challenge, a test in which neither the scientists nor the subjects know who is eating the potentially allergenic food and who is eating the harmless placebo.

What's missing in this battery of tests is a good animal model—a way of testing a new food on an animal first, instead of asking humans to become guinea pigs. Several animals have been tried: the guinea pig, of course, but also the rat, pig, dog, and mouse. None, as yet, has been shown to react to allergens exactly like a person would. As the University of Nebraska's Steve Taylor noted in an address published in *Nutrition Reviews,* "I would indicate to those of you interested in pursuing careers in this field that one of the most desperately needed things is a validated animal model."

The questions to ask to learn if a new protein will cause food allergies were published by WHO and the FAO in 2000 and 2001. The consensus of the American Medical Association is that by following these rules, "the overall risks of introducing an allergen into the food supply are believed to be similar to or less than that associated with conventional breeding methods." Samuel B. Lehrer, a professor of medicine at Tulane University and an expert on allergies to shrimp and corn, is more emphatic: "The allergy tests are so extensive that most of our foodstuffs would never pass them," he wrote.

※

When transferring any gene from a known allergen such as peanut, the FDA's 1992 guidelines considered it "prudent" for the plant breeder "to assume that the transferred protein is the allergen." The WHO-FAO report, written 10 years later, reinforces this suggestion: "It must be assumed that the novel gene product is allergenic unless proven otherwise."

Between these two recommendations, an allergen was indeed moved from one food plant into another. The project began because Pioneer Hi-Bred International wanted to increase the nutritional content of its soy-based animal feeds. Soybeans lack certain amino acids (methionine and cysteine), so supplements must be added to the feed for the animals fed on them to grow well. Knowing that the 2S albumin of Brazil nuts was a protein with a very high methionine content, Pioneer's scientists introduced a gene coding for the Brazil nut protein into soybeans.

At the time the 2S high-methionine albumin was not known to be an allergen. Studies done a few years previously in rats and mice had found it to be innocuous, although such studies could not give information about its allergenicity for humans. But because Brazil nuts were an allergenic food, Pioneer Hi-Bred asked a team of scientists—including allergy expert Steven Taylor—to test the new soybeans. When they were mixed with serum from people allergic to Brazil nuts, the IgE reaction was positive. Skin-prick tests were also positive. The 2S high-methionine albumin was a human allergen. Pioneer stopped development of the soybean, even though the product was intended for animals, not people. The soybean was never marketed.

Scientists, as well as consumers, have interpreted the results of this study in two ways. Marion Nestle, in a commentary that accompanied the team's report in the *New England Journal of Medicine*, said: "This study highlights gaps in our current knowledge of food allergies." For plant pathologist Jim Cook, on the other hand, "This is a perfect example of how the system works." In his testimony to Congress in 1999 Cook said, "It is always cited as how things can go wrong, but it is exactly how good testing in the laboratory can provide for safety."

Another example often cited as "how things go wrong" is the story of StarLink corn. Yet in this case there was no evidence—and still isn't—that the added protein was an allergen. A company called Aventis S.A., headquartered in Lyon, France, received approval in 1997 and 1998 to commercialize a variety of corn that contained both an herbicide-resistance gene and a Bt toxin gene. The approval was given by the EPA, whose primary concern is how a new crop will affect the environment. StarLink's application, for instance, had to explain how the new corn would affect honeybees, earthworms, minnows, mice, ducks, songbirds, and many other creatures, as well as other plants. Yet EPA also requires proof that any pesticide added to—or expressed in—a crop is safe to eat.

Bt-based pesticides have been used for more than 30 years to control a variety of insects, including gypsy moths. They are especially favored by organic farmers, who consider them natural, not synthetic. The toxins, which break down when exposed to sunlight, heat, or drying, come from a bacterium, *Bacillis thuringiensis.* "While commonly referred to in the singular as 'Bt,' *B. thuringiensis* is actually a large group of subspecies," writes entomologist Brian Federici of the University of California, Riverside. More than 70 subspecies (also called varieties or strains) have been identified. Each produces one or more types of "Cry" (for crystal-like) proteins in its spores. These proteins are not toxic until they come into contact with an insect's digestive juices, which are highly alkaline (with a pH reading of 8 to 10).

Eaten by an insect larva, the crystal dissolves. The protein is then broken apart, producing a toxic fragment. The fragment binds to a receptor on the lining of the insect's gut. If the insect doesn't have such a receptor—and most don't—nothing happens. In insects that have these receptors, on the other hand, the cell immediately begins to swell until it bursts. The active toxin binds to proteins on the epithelial cells lining the insect's midgut, forming pores that let potassium ions escape. Lacking potassium, the insect's gut cells take up too much water. Within two hours the insect stops feeding; if it has eaten enough of the toxin it becomes paralyzed and soon dies. The bacterium in this way prepares its own habitat. Federici points out that while Bt is usually thought of as a "soil bacterium," its true ecological niche is the dead insect. After the larva is dead, the bacterium feeds off its body, reproduces, and makes millions of spores.

The toxin has no effect on humans because of the differences between our digestive system and that of an insect. A human's digestive juices are highly acidic (with a pH reading of 1 to 3), so the crystal is not dissolved and the toxic fragment is not released. Even if it were, it would still not be toxic to a human. Toxins work by interfering with, or changing, a normal part of a cell. Human cells lack the receptor that interacts with the Cry protein in the insect's gut. For people, a Cry protein is just an extra bit of protein.

The genes that encode Cry proteins are on a large plasmid that *B. thuringiensis* strains carry. When the plasmid is lost—which can happen naturally or can be made to happen in the laboratory using heat—no toxins are made. Without the plasmid a Bt bacterium is indistinguishable from another species, *Bacillus cereus*. As Federici explains, molecular, biochemical, and physiological studies all agree that Bt and *B. cereus* are actually the same species: "The latter becomes the former when it acquires one or more plasmids that express genes for insecticidal proteins." Scientists retain the species name *B. thuringiensis*, though, for its practical use: it helps them classify the more than 20,000 isolates, individual Bt colonies that differ from each other in subtle ways.

More than 170 different Cry proteins have been identified in those 20,000 isolates; 16 of them have been approved by EPA for use in sprays and dusts or powders, of which 30 different formulations are sold. These are made by fermentation—growing bacteria in large vats and separating out the toxic spores, then adding stabilizing agents, spreaders, stickers, and fillers. To make sure they are safe to use, the EPA established a three-tier series of tests. First laboratory animals, including insects, birds, fish, and mammals, are exposed to a hundred times the amount of Bt spores recommended for use in a crop field as a pesticide. If there are any signs of toxicity, that Bt isolate is rejected. If the results are unclear, tier II tests are conducted, in which the animals are exposed multiple times. Tier III tests include two years of feeding trials. "To date," writes Federici, "*none* of the registered bacterial insecticides based on Bt have had to undergo tier II testing." As a result, no limit is set by the EPA for the amount of insecticide residue that can be left on a crop. Bt can be sprayed on tomatoes, for instance, just before they are picked and eaten—the EPA doesn't even require the vegetables to be washed before they are sold. Writes Federici, "It is important to realize that such a statement cannot be made for just about any synthetic insecticide."

Most of the insecticidal sprays and powders contain four Bt toxins to control a wider range of pests, because each Cry protein is toxic to only some kinds of insects. Some affect gypsy moths, mosquitoes, blackflies, cabbage loopers, and boll weevils. Others—particularly the ones named Cry1Ab, Cry1Ac, and Cry9C—kill the European corn borer, the number 2 insect enemy of corn farmers. In bad years the pest has cost farmers an estimated $1 billion, or about 20 bushels per acre.

Several varieties of corn have been genetically engineered to incorporate a Bt gene. Most produce Cry1Ab. StarLink corn was the only one to use Cry9C. In spite of the fact that Bt was a familiar pesticide whose action was well known and whose toxicity was well tested, the EPA did not simply assume that Cry1Ab and Cry9c, when introduced into corn, would not be toxic, nor that they were free of allergenic potential. Both proteins were put through tests. Cry1Ab was shown to be easily digested. It was inactivated by cooking. Even at relatively high doses (4,000 milligrams per kilogram), when fed to a laboratory animal it was not toxic. Corn containing Cry1Ab was approved for human use.

StarLink corn, however, was approved only for animal feed. Cry9C did not break down as quickly as Cry1Ab when it was mixed with stomach acid and digestive enzymes. Such resistance to digestion is one of the criteria used to identify potential allergens. This result didn't mean that Cry9C is necessarily allergenic, only that it was suspect. The EPA required further testing before StarLink corn could be allowed in human foods. Aventis CropScience, the company's American subsidiary, decided to commercialize its variety as animal feed first, then follow up with more testing and a later petition for it to be used in human foods.

Farmers who bought StarLink seeds were supposed to keep their corn segregated from other varieties by planting a 660-foot buffer strip of another crop between it and any other type of corn. They were also not to sell it for use in human food. Yet most field corn—the kind used for animal feed, as well as chips and taco shells or even gasohol—is stored by grain dealers in the same huge silo-like grain elevators. On a practical basis, the promise Aventis sought to extract was very hard for a farmer to keep. In the opinion of the attorney general of Iowa, it was "irresponsible" for Aventis to market its seed corn with such unrealistic restrictions. He did not believe that most Iowa growers even knew about the rules. An Aventis-sponsored survey of 230 farmers in December 1999, in fact, found 2 farmers who had knowingly sold their harvest for use in human food. Another 29 did not know what their crop had been used for. Aventis did—responsibly—report these results to the EPA, but no action was taken. A few months later a panel of scientists was convened to recommend a way to test StarLink further, in hopes of approving it for human consumption.

Before their suggestions could be acted upon, the *Washington Post* reported that a company called Genetic ID had detected the presence of StarLink corn in Kraft Food's Taco Bell taco shells. As Genetic ID explained, the environmental group Friends of the Earth had contracted with it to test the taco shells. The company claimed to detect the 1 percent StarLink corn in the taco shells using a PCR test that it has not disclosed. Genetic ID's president, John Fagan, is the author of the book, *Genetic Engineering: The Hazards, Vedic Engineering: The Solutions: Health—Agriculture—The Environment*. He was one of 26 plaintiffs who sued the FDA in 1998 for inadequate testing of foods modified using biotechnology. Their suit was dismissed by the United States District Court for the District of Columbia on September 29, 2000, 11 days after the StarLink story broke.

The publicity triggered a massive recall of taco shells, corn chips, cornmeal, and other foods containing corn. Aventis offered to purchase the entire 2000 StarLink harvest and stopped all sales of StarLink seed corn. The USDA published a recommended method to distinguish StarLink from other corn, and grain elevators and seed companies started testing their stores. The tests—which were sensitive enough to find one StarLink kernel in a sample of 400 kernels—turned up much more StarLink DNA than expected. StarLink had been grown on only 0.5 percent of all the corn acreage in America in 2000. Yet corn varieties that were not supposed to contain the gene tested positive. Even products made from white corn—StarLink was a yellow variety—tested positive. Farmers in 17 states whose nontransgenic products tested positive threatened to sue. They were backed by their state attorneys general, who negotiated an agreement with Aventis to reimburse growers and grain elevator operators who had suffered losses. Aventis would pay 25 cents per bushel to farmers who grew StarLink, to those who grew other varieties of corn within 660 feet of a StarLink field, and to those whose corn tested positive for Cry9C. The company also agreed to cover losses from transportation, storage, and testing. Aventis expected the settlement would cost it at least $100 million. The top leadership of Aventis in the United States—president, general counsel, and vice president for market development—were fired. The U.S. Department of Agriculture offered another $15 to $20 million to compensate small seed companies whose stocks were contaminated with Cry9C. By 2002 all StarLink corn was eliminated from the U.S. grain supply.

Meanwhile, calls began coming in to the FDA reporting allergic reactions to taco shells and corn chips. As these calls accumulated, the FDA requested that the Centers for Disease Control (CDC) test the people reporting such reactions to see if their blood serum contained the IgE sensitive to Cry9C, the purported culprit. The CDC determined that 28 of the people who had filed complaints had indeed experienced allergic reactions—but none of them had a form of IgE that would react to the Cry9C protein. They were allergic to something else. This result is not altogether surprising given what is known about allergic reactions. An IgE molecule that matches and binds to Cry9C can be made only after the body has been exposed to Cry9C—repeatedly, and in significant amounts. Because Cry9C had not been in foods before, allergy expert Steven Taylor was quoted as saying, "There is virtually no risk associated with the ingestion of StarLink corn in this situation."

In July 2001, an EPA committee reviewed the StarLink case. In petitioning for approval for StarLink to be used in human food, Aventis had submitted data that showed that the Cry9C protein broke down more easily when it was wet or heated—both essential steps in making taco shells. The EPA's scientists confirmed that, indeed, the normal process of preparing corn for that use, called wet milling, removed almost all of the Cry9C. The question of whether that small amount was harmful to health went unanswered, because, as the EPA had reported several months earlier, "Aventis requested voluntary cancellation of their Cry9C StarLink corn registration and this cancellation became effective on February 20, 2001."

The tests sponsored by Friends of the Earth, and the subsequent publicity and recalls, did not prove that StarLink corn contained an allergen. They proved only that the attempt to separate corn for human food from corn for animal feed had failed. For agreeing to the EPA's request not to sell StarLink for use in human food until more data were submitted, Aventis lost more than $100 million in cash and an uncalculated amount in farmer confidence and consumer goodwill. Bayer bought Aventis CropScience in 2001; the company is now called Bayer CropScience.

StarLink corn, the Brazil nut-soy case, and the poisoned rat debate have all received worldwide publicity. Each has been transformed and sensationalized with retelling. StarLink corn and the Brazil nut protein-containing soybean have been credited with causing allergic reactions. Genetically modified potatoes have been accused of poisoning rats. But the facts don't support the stories. StarLink caused no allergic reactions, the soybean didn't reach the market, and the new genes in the potatoes were never shown to have caused harm. However titillating, might this sensationalism itself not be the poison?

What is rarely mentioned in debates about genetically modified food is how it has made our food *safer.*

Entomologist Brian Federici writes, for instance, "One notable improvement in safety that has not received much attention is that Bt maize grain contains lower levels of certain fungal toxins known as fumonisins. These toxins can cause illness and death in horses and pigs, and have been implicated in certain forms of liver and esophageal cancer in humans." When the European corn borer tunnels into an ear of corn, it opens a passage for *Fusarium* fungus. The fungus is not easy to see, and infected ears are harvested along with their undamaged neighbors. The toxins have been found in corn around the world. They are especially a problem in countries with warm or subtropical climates. A study in 2000 found that 82 percent of the corn grown in the islands of the central and south Pacific, 77 percent of the corn in Africa, and 63 percent of the corn in North and South America was contaminated with fumonisins. In corn meant for use as human food, the levels of toxins ranged from one milligram per kilogram up to 150 milligrams per kilogram. Horses die of brain damage when they are fed infected corn containing more than 10 milligrams of toxin per kilogram of corn. Rats fed corn tainted with 150 milligrams of toxin per kilogram suffer liver and kidney cancer.

One way to stop the toxins from accumulating in the corn is to limit the holes through which the fungus can get into the plant to multiply. By killing the European corn borer as soon as it starts eating, StarLink corn and the other varieties engineered to produce Bt toxins do just that. Comparisons of Bt corn and conventional varieties in Iowa found that the Bt corn contained 93 percent *less* of the toxic fumonisins. Similar results have been reported in France, Italy, and Spain.

Bob Buchanan, a plant pathologist at the University of California, Berkeley, happened onto another way in which genetic engineering can make food safer. Thirty years ago he began studying a small protein called thioredoxin. First discovered in bacteria, it was found to play a part in regulating photosynthesis. Buchanan's team discovered that thioredoxin somehow alleviates food allergies, as well as enhancing the digestibility, nutritional quality, and baking quality of flour. Working with Buchanan are allergy specialist Oscar Frick of the University of California, San Francisco, and geneticist Peggy Lemaux of U.C. Berkeley. As Buchanan said in 1999, "We seek to gain an understanding as to how thioredoxin from our laboratory, when engineered into cereals by Dr. Lemaux's group, could benefit humans. In addition to barley and wheat, our agenda includes rice, corn, sorghum, soybean, and peanuts."

Other researchers are addressing the allergy issue a different way. Samuel Lehrer at Tulane University is identifying the parts of a protein molecule that allow the immune system's gatekeeper, IgE, to bind to it and set off the cascade of symptoms we call an allergic reaction. "If you can substitute one amino acid in the protein," he explains, "you can completely abolish IgE binding." Lehrer's group is working with shrimp, while another team at the University of Arkansas and Arkansas Children's Hospital in Little Rock is concentrating on soybeans and peanuts. "The group in Little Rock are farther along," he reported at the American Association for the Advancement of Science's annual conference in 2003. "They can almost completely abolish IgE binding. There are still serious issues to address, though," he added. "We need to demonstrate that these altered proteins don't contain *new* binding sites." Lehrer concluded, "It is our hope that in the not-too-distant future foods can be developed that will substantially reduce the number and severity of food-induced allergic reactions."

Eliot Herman of the Donald Danforth Plant Science Center in St. Louis has also been trying to breed a less allergenic variety of soybean. He has targeted an enzyme called P34, which is the major allergen in soybeans. In the plant's leaves and stems it seems to help fight off diseases. Unfortunately for people, it accumulates in the beans. Herman's modified soybeans, now in field trials, completely lack P34. The gene that directs the production of the protein in the seed has been silenced. Nothing else about the plant has been changed. Herman writes, "These results show that it is feasible to use biotechnology to suppress a major human allergen in crops, implying that widespread use of this approach can improve the safety of food for sensitive people."

10 THE BUTTERFLY AND THE CORN BORER

Because science and politics have become so entangled, it has been hard to pick rationally through the facts.

—Michael Specter (2000)

Victor Davis Hanson is a fifth-generation grape farmer and orchardist in the San Joaquin Valley of California. He is also a professor of classics at California State University. In *The Land Was Everything: Letters from an American Farmer*, he writes eloquently about "a race that we cannot win, between land farmed and land paved. Our ancestors," he notes, "beat back the wild to give us these farms—we of a lesser breed were beaten back by new men to give them up. We battle the encroaching suburb; our grandfathers in their youth once fought hunger and typhoid—both of us have received four dollars a box for plums."

Asphalt is, of course, not his only enemy, nor the one he spends his days battling. "People must realize," he writes, "that farming is not statistics from the Department of Agriculture nor quotes from some scrubbed agribusiness talking head, but rather is a very unpleasant and brutal task to bring food out of the dirt." Against the "flyers, hoppers, and crawlers," for instance, those "pests of biblical magnitude," he sprays poison. "The trick," he says, "is merely one of logistics, not of poison per se: providing the dose in enough quantity and strength to kill the swarm before it demolishes the vineyard in question."

Farmers have long had first-hand knowledge that they were also, while saving the crop, poisoning other creatures who did no harm. Hanson writes, "Organosphosphates killed every living thing in the orchard and the vineyard: mites, hoppers, and worms, but also predator spiders, wasps, and ladybugs, even birds, gophers and snakes—and the occasional laborer. Even the lowly toad snug in his hole beneath the vine was not safe. I have seen these cute amphibians still and yellow outside their holes on the day after the spray."

Still, Hanson sprays.

"Why, you ask, would any sane man, as I have, apply even the less toxic ancestors of parathion—Lannnate, Guthion, dimethoate, lindane, carbaryl, Thiodan, or diazon—to kill everything on leaf and tendril?" Because the alternative is to cash it in and let the "new men" pave over the farm.

Not that Hanson hasn't tried less toxic means. Not that he doesn't wish, fervently, for a better way. A farmer's glossary, he says, defines a spray rig as, "a mechanical Faust that can save your crop and you along with it, at the price of your soul."

Referring to a brand of pesticidal Bt, he writes:

> If you dust the organic bacteria Dipel on grape leaves, it will deter, but not stop, a grape-leaf skeletonizer or coddling moth. If, as one should, you sow cover crop, plant berries and flowers, and use manure, you can grow predators to chomp on, but not eradicate, deleterious insects. Is it to be hours of contemplation, repeated application, and constant monitoring, or one shot of caustic Omite that will kill red spider mite for two months? The choice is not really the blinkered farmer's alone. The decision has already been made by the environmentally correct consumer: He wants fruit plentiful, colorful, hard, and fresh, free from crack and scar—not high-priced and tasty, overripe, leaking, and pockmarked, a hitchhiking gnat or stowaway fly now swirling amid the glistening produce section at the local supermarket. Under no condition must the vine-hopper's nontoxic excrement dot the grape. An English professor once called me complaining about the grapes he bought at the local organic farmer's market on campus. "They have black dots all over the bunches," he raved.
>
> "That's hopper crap," I replied.
>
> "Well, who wants to eat it? Not me," he fumed.

No grapes have been genetically modified to carry their own insecticide (although researchers in California are developing a grape that is protected against a bacterial disease). Yet the crops that produce toxic Cry proteins, using a gene from the bacterium *Bacillus thuringiensis,* were designed with Hanson's concerns in mind. Bt crops are intended to keep "non-target species," from the lowly toad to the laborer, safe. The first Bt gene was cloned and transferred to *E. coli* in the early 1980s. By 1990, with some adjusting of the DNA coding sequences to reflect differences in the abundance of certain kinds of RNA in plants as compared to bacteria, tomato, tobacco, and cotton had been modified to make their own Bt toxins. Bt corn and potatoes were reported in 1993. In field tests it was found that the toxic Cry proteins worked. The only creatures killed by the poison were the ones that ate the crop—which is a fair definition of "pest."

It is ironic, then, that the fight against genetically modified foods was inflamed by a scientific finding that pollen from Bt corn could, in the laboratory, kill the larva of a monarch butterfly.

Monarch butterflies have been called "the great fluttering pandas of the insect world." Like pandas, eagles, wolves, whales, elephants, snow leopards, baby seals, and other charismatic mega-fauna, monarchs are much beloved and much photographed. They are emblematic in many people's minds of the word "butterfly." As well as being wonderfully picturesque in their bright orange and black habit, monarch butterflies are also mysterious: they undertake a heroic migration each year from Canada and the United States to a patch of Mexican forest, there to cluster in brilliant bunches in the trees.

How they navigate over such distances, and why millions congregate in so small a space are unknown to science. Much of their lifestyle, though, is plain. Monarch butterflies depend on the milkweed plant; they are also known as milkweed butterflies. They lay their eggs on the undersides of the weed's upper leaves and, when the larvae hatch, the caterpillars eat the milkweed leaves. They eat only milkweed leaves for two weeks, growing and molting several times. Then, when the caterpillar is about 2 inches long, it spins a cocoon, in which it is transformed into the beautiful butterfly—the beautiful and noxious butterfly. For the milkweed plant is poisonous. The monarch larvae accumulate the toxin from the leaves they eat, and the poisons persist in the butterflies. These toxins make the monarch repulsive to most birds and other predators. A bird will bite a monarch once and drop it, never to bother such an insect again. The ploy is so successful that other butterflies, which evolved into monarch look-alikes, also avoid being eaten, even though they are not themselves toxic.

In May 1999, in a one-page letter published in the journal *Nature*, John Losey and two colleagues in the entomology department of Cornell University told the world that pollen from Bt corn plants kills the larvae of the monarch butterfly. Losey's paper was front-page news in the *New York Times*; the monarch photo accompanying the story was captioned "Bambi of the insect world." The news traveled around the world, from the BBC to the *Boston Globe* to the *San Francisco Chronicle*.

The study immediately turned up the volume of sentiment against genetically modified foods. Greenpeace members in butterfly costumes mimed being "felled by killer corn." Friends of the Earth, soliciting new memberships, sent a letter asking, "If deadly toxins that kill butterflies are being introduced into our food supply, what effect are these toxins having on you and your family?" The activism itself fueled further coverage in the news, and calculatedly so. Margaret Mellon of the Union of Concerned Scientists recently admitted to a journalist that "We worked hard to make this a high-profile issue because without media attention we knew nothing would be done. We saw the findings as an illustration of how superficial risk assessment for GM foods was."

Inside the scientific community, Losey and his colleagues were criticized for overstating their case. Their laboratory study was imprecise and—more importantly—seemed far from the conditions a monarch would encounter in the wild. Losey's team had sprinkled either Bt corn pollen or regular corn pollen onto an array of damp milkweed leaves. Onto each leaf they set five three-day-old caterpillars. They repeated the experiment four times. Of the larvae exposed to Bt corn pollen, 44 percent died; those that lived were smaller than the larvae in the control group. The researchers concluded that Bt corn had "potentially profound implications for the conservation of monarch butterflies."

"Never mind," reporter Bill Lambrecht wrote, "that just eleven caterpillars died. Or that the pollen might not have been spread uniformly on the leaves and that they had run out of the pollen, preventing them from broadening the scope of the project." Losey had not quantified the amount of pollen on each leaf, nor had he given the caterpillars a choice between plain milkweed leaves and those with the presumably untasty pollen on them.

The Cry protein to be expressed in corn was chosen specifically because it did kill caterpillars. To entomologist Anthony Shelton, also at Cornell University, Losey had no story. "Every entomologist knows," said Shelton, "if you feed monarch butterfly larvae Bt toxin, whether it be in corn or whether it be on a spray, that insect will die." Many environmentalists know it as well. They have used the same argument to fight the widespread spraying of forests to kill the alien gypsy moth. Bt is the pesticide of choice to control this voracious caterpillar, which will defoliate whole forests if left unchecked. Spraying Bt from helicopters saves the trees, yet kills most other butterflies and moths, including monarchs. In the case of Bt corn, choosing between the risks to the butterfly and the benefits to people seems far easier. As John Foster, an entomologist at the University of Nebraska, wrote: "There probably was not an entomologist in the world who was not aware that corn pollen containing the Bt gene could harm butterflies—if butterflies ate corn pollen, which they don't."

Monarch caterpillars in the wild eat milkweed and only milkweed. In their butterfly stage they sip nectar. The only way corn pollen can harm them is if it falls on a milkweed leaf while they are hungry caterpillars and they inadvertently eat it. Milkweed grows both in and outside of cornfields. Inside the field, it is a weed—one the farmer diligently tries to eradicate. If outside the field, it still needs to be near enough that the heavy corn pollen can reach it in quantity. The two events also have to be synchronized. Losey's study didn't mention the fact that corn pollen had to be shed and fall on the milkweed leaves at precisely the same time the butterflies were laying their eggs and caterpillars were hatching.

Because of the unknowns—and the publicity—a remarkable study was launched. It was funded jointly by a pooled grant provided by the Agricultural Research Service of the USDA and the Agricultural Biotechnology Stewardship Technical Committee (whose members were Aventis CropScience, Dow AgroSciences, E. I. du Pont de Nemours and Company, Monsanto, and Syngenta Seeds, Inc.), together with funding provided from the Canadian Food Inspection Agency, Environment Canada, the Ontario Ministry of Agriculture, Food, and Rural Affairs, the Maryland Agricultural Experiment Station, and the Leopold Center for Sustainable Agriculture.

Researchers from the USDA and several universities looked at all the factors likely to determine the effects of Bt corn pollen on the larvae of monarch butterflies. They performed tests in the laboratory and in actual cornfields. One group measured exactly how toxic the various Cry proteins were to caterpillars, as well as the amount of toxin present in the pollen grains of three widely used Bt corn varieties. Other groups measured how much milkweed grew in, near, and far from cornfields at sites from Ontario west to Wisconsin and Iowa, and down through New York to Maryland. Others counted the number of pollen grains deposited per square centimeter of leaf surface on milkweed plants in cornfields, near cornfields, and far from cornfields. Did the upper leaves of milkweed plants, on which the larvae fed, they asked, receive the same amount of pollen as leaves farther down? How was the pollen density affected by wind and rain? Still others asked whether monarchs laid more eggs on milkweed plants inside cornfields, where milkweed is a weed, or on plants outside cornfields. Finally, they determined the extent to which egg laying coincided with the week to 10 days during which corn plants shed their pollen.

The results of all these studies—including one co-authored by John Losey himself—were published together in September 2001. The timing could not have been worse. "The world might have a different attitude to the safety of genetically modified (GM) crops if it had not been for the terrorist attacks on New York and Washington," wrote the editors of SciDev.Net in June 2002. SciDev.Net is an Internet news service sponsored by the journals *Nature* and *Science* in association with the Third World Academy of Sciences and is funded by the United Kingdom, Canada, and Sweden. "For among the many stories that these attacks obliterated from the newspapers," the editorial continued, "was a report on six scientific papers published simultaneously in the *Proceedings of the National Academy of Sciences (PNAS)* on the safety of corn that has been genetically engineered to express a toxin that kills certain pests."

There were no Greenpeace costumers, no Friends of the Earth letter-writers to ensure that the news reached page one under a photo of the "Bambi of the insect world." There was no effort by the Union of Concerned Scientists to make it a high-profile issue, as there had been when Losey's original study was published. Consequently, noted SciDev.Net's editors, "the image of dead monarch butterflies lingers in the public consciousness as the unacceptable face of GM technology."

A report by the Pew Initiative on Food and Biotechnology notes that Losey's initial study and the "firestorm that it unleashed" had good consequences. The EPA made some intelligent changes in its regulatory process. And when the registrations on five Bt-corn varieties came up for renewal in October 2001, "The decision to go ahead was taken after a vigorous debate in which all stakeholders—from industry representatives to environmentalist groups—had been given a chance to examine the Cornell research and its implications in detail, to assess the results of follow-up studies, and to talk through their differences," according to SciDev.Net. The result, they reported, was "a robust social consensus between all sides of the debate that had previously been lacking."

News of this robust social consensus did not make it to page one of the world's newspapers. In 2003, when *Prospect Magazine* in the U.K. printed "The Butterfly Flap," journalist Peter Pringle instead summarized the results of the "vigorous debate" as: "Losey and co. had lost a battle." The *PNAS* series did uncover some unappreciated facts about monarch butterfly biology. While corn pollen is shed well before the peak of monarch egg laying in Maryland, the two events overlap in more northern areas, such as Minnesota and Ontario. In Ontario, most of the milkweed that supports monarch caterpillars is outside cornfields. In other places, particularly in the Midwest Corn Belt, monarchs laid their eggs on milkweed plants inside cornfields just as often as they did on milkweed outside cornfields. The studies also confirmed, as plant breeders had argued, that corn pollen is rather heavy and most of it comes to rest inside the cornfield. A few steps away from the field (2-3 yards), only a fifth as much pollen landed as did inside the field.

As for the risk of Bt corn, the studies determined that only one of the three commercial varieties produced enough Cry protein to kill monarch caterpillars—or even to slow their growth—at the actual pollen densities detected in cornfields. At the time of the study, this variety—called event 176 hybrid and sold by Novartis under the name KnockOut and by Mycogen as NatureGard—accounted for only 2 percent of the corn planted. It had already been slated for phaseout. According to Chuck Armstrong of Monsanto, "The product concept was to express Bt in the foliage and the pollen. But insects still got inside the plant and caused the damage. Farmers found one kind of Bt corn, Yieldgard, worked well, and Bt176 didn't work very well. The monarch butterfly study was just one more nail in the coffin. It wasn't *the* reason event 176 went off the market." In 2001 the registration for event 176 hybrid was allowed to lapse, and it is no longer sold.

In a summary paper in *PNAS*, Mark Sears of the University of Guelph in Canada and several other authors calculated the worst-case scenarios for Iowa, a major corn-producing state. At one extreme, they assumed that KnockOut and NatureGard would increase to 5 percent of the area planted. When they combined this assumption with the amount of egg-laying-pollen-shed overlap, the actual pollen densities, and the proportions of milkweed both inside and outside cornfields, they found that about 0.4 percent of the monarch population (or 1 monarch out of every 250) would be at risk of exposure to toxic effects from Bt corn.

A second scenario assumed, instead, that KnockOut and NatureGard corn were phased out (as indeed has happened) and the pollen of future Bt varieties would be only as toxic as the two other varieties tested (Syngenta's Bt11 YieldGard or Attribute, and Monsanto's Mon810 Yieldgard). Then, even if 80 percent of the corn acreage in Iowa were planted in Bt varieties—and this is the maximum expected if the current stipulation for Bt-free refuges, or refugia, is maintained—only 0.05 percent of the state's monarch population might be in danger. Five hundredths of a percent means one larva out of every 2000.

Per Pinstrup-Andersen, then director general of the International Food Policy Research Institute, summed up the monarch butterfly fracas in 2000: "The attention given to this particular case contrasts strangely with the general attitude toward standard agricultural practices: conventional pesticides, which do not differentiate between friend and foe, kill a great many harmless insects. The fierce denunciation of clearly targeted and contained pest control through Bt plants is perhaps a little out of touch with reality."

The Associated Press, to its credit, did try to include a reality check in one news story in 1999. It contacted Jeffery Glassberg, president of the North American Butterfly Association, and asked his opinion of the threat of Bt corn. Said Glassberg, "I think there are a lot more dire threats than that to monarchs. In the Midwest, mowing roadsides and using herbicides is probably much more devastating, actually."

Anthony Shelton, Losey's colleague in Cornell's entomology department, is equally concerned about the side effects of Bt crops. He does not worry about the monarch: "How many monarchs get killed on the windshield of a car?" was his flippant answer to one journalist who pressed the point. But he has warned since 1995, when the first Bt plant (cotton) was commercialized, that the crops' success could very well spell their failure—and that of Dipel and the other Bt sprays on which organic farmers depend.

Plants and insects, as entomologists well know, wage ongoing chemical battles for survival. Like the milkweed, plants produce toxic chemicals in an effort to keep insects from eating them. Like the monarch, insects evolve ways to evade the chemical tricks of plants. The interests of humans are generally aligned with those of the plants—we don't want bugs getting to our food first—with one rather large exception. The chemicals that plants make to protect themselves sometimes irritate us as well. The lectins in kidney beans are there to ward off insects, as are the glycoalkaloids in potatoes and cassava, and the psoralens in celery. It is no accident that domesticated crops are less toxic, in general, than their wild ancestors. As Michael W. Pariza, a specialist on cancer risks in food at the University of Wisconsin-Madison, has noted, "In this context, crop breeding can be seen as a means of controlling unwanted wild traits, such as toxin production, in plants destined for the table."

"Unfortunately," Pariza adds, "selected traits that make a plant desirable as human food also make it desirable to insects. This, in fact, is why synthetic pesticides are used: they replace the naturally occurring pesticides and related survival traits that have been bred out of food plants."

Crop breeding has, at times, worked the other way. Jack Harlan, the expert on crop evolution, points out that many Indian tribes prefer to grow highly toxic varieties of cassava. The poisonous varieties, while requiring extensive care when preparing and cooking, were less damaged by insects and disease, and so produced a greater yield. The developers of the Lenape potato, in the 1960s, also selected for high toxicity—without meaning to—while trying to breed in disease and insect resistance. Again, in the 1980s, trying to reduce the need for chemical insecticides, crop breeders used traditional breeding techniques to create a celery that naturally warded off its pests. Ordinary celery contains about 0.8 parts per million of psoralen, the plant's natural pesticide. The new version contained 6.2 parts per million, more than seven times as much. Bugs did not like it a bit. But neither did the celery pickers, on whose hands and arms it caused rashes. The new celery was quickly withdrawn from the market. Later research found that the psoralen in celery was also carcinogenic when fed to rats and that an ordinary celery plant, if under stress from drought or insect damage, will increase its production of the toxin to 25 parts per million, 30 times the normal level.

Chemical insecticides are, from an insect's point of view, no different from the plant's own toxins. And their widespread use in modern agriculture has had a highly predictable result: the emergence of insects that are resistant to certain pesticides. Sprayed, they do not die. Sprayed with a double dose, they still might not die. They have won that particular chemical battle. How? Through mutation. Although a mutation in just the right gene to make an insect proof against a toxin is very rare, the number of insects in nature is very large—so large that the heavy use of insecticides rapidly selects for the ones with the right kinds of mutations. All the others die, and only the few that are resistant reproduce and multiply. But, as Victor Davis Hanson has seen in his vineyards, those few are enough. "Mites and hoppers breed geometrically, not arithmetically," he writes in *The Land Was Everything*. "Their reproductive cycles overlap, as populations go from hundreds to thousands to millions to billions in a vineyard in only days." Again, Hanson and his fellow farmers must roll out their Faustian spray rigs. Only this time the poison must be different, for these billions of bugs are resistant to the old one.

Bt crops, in theory, can make farming safer for farmers (and, incidentally, for butterflies). Because the Bt genes protect plants from within, the use of the spray rig, and the accidents that accompany that use, should be less. But farm safety is not, in fact, behind the spectacular success of Bt corn. As professors at Colorado State University explain in their handbook on transgenic crops, "Field corn is not usually sprayed with insecticides because there is some market tolerance for insect damage on this kind of corn." It is either fed to animals or ground and processed into corn chips, cornflakes, and cornmeal, so wormholes and other blemishes are not generally noticed. Worm-eaten sweet corn, on the other hand, is, like black-spotted grapes, hard to sell. "Sweet corn is sprayed with insecticides frequently, sometimes every two or three days, to ensure that the ears will be attractive at harvest." But while Bt sweet corn has been approved by the EPA, it is not being grown.

Instead of replacing the chemical pesticides, Bt field corn has shown farmers the true cost of the damage done by the European corn borer. In its moth stage, the corn borer lays eggs on young corn plants, where the larvae hatch. They feed briefly on the leaves, then, as their name implies, bore into the cornstalk. The damage their tunnels do goes unseen and unchecked until, in a heavy rain or wind, the riddled cornstalk topples. Unless a farmer scouts his fields often, an infestation is easily overlooked until too late. Even the efforts of the 5 to 8 percent of farmers who did try to spray against corn borers were often without effect. Once the worm is in the corn, the poison cannot reach it. The corn borer can produce two generations in each crop cycle; in the South, it reaches three generations, or even four. The only sure way to control it is to stop planting corn altogether.

Bt corn first went on sale in 1996; as of June 2002 more than a third of the field corn in the U.S. (as well as more than 70 percent of the cotton) were Bt crops. Bt varieties were sold not only by Monsanto, but by Syngenta, Aventis, Mycogen (owned by Dow AgroSciences), and Pioneer Hi-Bred (owned by duPont). Bt corn did not immediately reduce farmers' pesticide use. As the 2000 report to Congress on the benefits, safety, and oversight of agricultural biotechnology noted, this fact "is often cited by critics of biotechnology as an example of a bioengineered crop that has not met expectations." Rebecca Goldburg of the Environmental Defense Fund, for example, complained to Congress that "Bt corn largely supplements rather than substitutes for insecticide use on field corn." And yet few farmers had expected Bt corn to cut their use of sprays. According to a study by Iowa State University in 1998, 82 percent of the farmers in the Midwest who had planted Bt corn that year said their primary reason for doing so was to prevent losses from the corn borer. Those losses, according to the National Center for Food and Agricultural Policy, a nonprofit research organization, were equal to 3.5 billion pounds of corn by 2001, an amount worth $125 million.

Now that the enemy has shown its face, farmers are unlikely to ignore it once again if, for any reason, Bt corn becomes unavailable. Having seen how much more corn each acre yields when corn borers are eliminated, the National Center for Food and Agricultural Policy study suggests, farmers "would be likely to take the pest more seriously and apply insecticides more frequently than in prior years." In a typical year, the study calculates, farmers would spray an additional 2.6 million pounds of pesticide to kill corn borers if Bt corn were not available.

Why would Bt corn become unavailable? Because of its spectacular success, Shelton and other entomologists have argued. Just like a chemical insecticide, Bt kills sensitive insects. Those insects with rare mutations that make them resistant to the toxin will be a larger fraction of the survivors. With the introduction of Bt genes into so many different kinds of plants (by 1999 18 crop varieties had been approved for field testing, and 16 companies were at work developing more), and with the acreage being planted in Bt crops expanding so rapidly, entomologists feared that Bt crops would select for resistance faster than anyone anticipated. An early estimate was that insects might become resistant in no more than five to seven years—or even less.

Because there are more than 170 different toxic Cry proteins, some scientists suggest that saying insects will become resistant to Bt is like crying wolf. Biochemist Milton Gordon of the University of Washington, for instance, writes that talking about Bt toxin "as a single compound is very similar to talking about all of the antibiotics that have been discovered and are now being used in humans as a single compound. If the pathogenic bacteria become resistant to one type of antibiotic, it is possible to switch to another type and still get good results. The same is true of Bt."

And yet for the company that has developed a Bt corn variety, Gordon's argument is little solace. Once insects become resistant to its toxin, a Bt crop would have no advantage over another, less expensive variety; both would need to be sprayed to protect them against the corn borer. Because developing and testing new Bt crops is fiercely expensive, it was important for companies to invest in strategies that would extend the useful lifespan of each Bt variety.

The EPA was also concerned. Part of its mission is to limit the environmental risks of pesticides and to promote safer means of pest management. Bt, used by home gardeners and organic farmers for decades, has a reputation for being a benign and environmentally friendly insecticide.

That reputation might be undeserved. *Bacillus thuringiensis* differs from another soil bacterium, *Bacillus cereus*, only in that *B. thuringiensis* contains a plasmid, an extra ring of DNA, that gives it the ability to make toxic Cry proteins, and *B. cereus* does not. Otherwise the two species are indistinguishable. A team of microbiologists from the University of Oslo who studied the genes of the two bacteria argued in a paper published in *Applied Environmental Microbiology* in June 2000 that they were, genetically, the same species despite the different names. The name scheme matters, they pointed out, because *B. cereus* is "an opportunistic pathogen that is a common cause of food poisoning." The researchers found that a third *Bacillus* species should also be lumped together with *B. cereus* and *B. thuringiensis*: the third species is *Bacillus anthracis*. "*B. anthracis*," they write, "causes the acute fatal disease anthrax and is a potential biological weapon due to its high toxicity." But for that one plasmid, the Bt whose spores are sprayed on organic vegetables is the same as anthrax. The plasmid is known to move into and out of Bt quite easily. The fact that the three bacteria belong to the same species is not just a matter of scientific interest: genes, including plasmids, are naturally shared among members of the same species.

That said, Bt remains the safest insecticide on the market, one which farmers are not even required to wash off of their produce before selling it. The loss of one Bt toxin, like the loss of one antibiotic, might not matter. Yet many of the Cry proteins engineered into plants are also found in Bt sprays. A bug that became resistant to the plant would also be unharmed by the spray. Consequently, a 2001 EPA report noted that the loss of a popular Bt toxin, if insects did become resistant to it, could have "serious adverse consequences for the environment": conventional growers would have to shift to more-toxic pesticides and organic farmers would lose a valuable tool. Keeping pests susceptible to Bt, the EPA decided, was "in the public good."

"This issue is not new to agriculture," plant pathologist Jim Cook explains. "Resistance breeding is an ongoing effort for crops just to stay ahead of the ever-evolving populations of pest species."

In *The Living Fields,* Jack Harlan illustrates this effort to stay ahead in the battle between wheat and the fungus *Puccinia,* which causes the disease called rust. "In North America," he writes, "we managed to plant a carpet of wheat from northern Mexico well into the prairie provinces of Canada." Although patches of other crops are mixed in, none is large enough to keep the rust spores from traveling on the wind from one wheat field to the next over a total distance of some 2,500 miles. Each year the disease hits first in the south, along the Rio Grande valley. The fungus germinates. Its thread-like mycelia consume the wheat's stems and leaves, then set loose on the wind great numbers of spores. Some of these blow northward to the next wheat field, where their life cycle starts over. "The rusts cannot overwinter in the northern end of the wheat belt and must arrive each year from the south if wheat is to be rusted," Harlan explains. Nor can the fungus withstand the summer heat in the south, "so spores must be blown southward to infect winter wheat and renew the spring migration." Our 2,500-mile carpet of wheat allows this yearly migration to happen. Says Harlan, "The system must be included among the marvels of the biological world."

A Canadian-bred wheat called Marquis was planted across the northern plains in 1912 because rust did not harm it. "The rust epidemic of 1916 took it out," Harlan says. A derivative of Marquis called Ceres was the favorite from 1926 to 1934. "Severe epidemics in 1935 and 1937 took out Ceres. The epidemic of 1935 is estimated to have cost 160,000,000 bushels in the USA alone." Two other varieties, Thatcher and the aptly named Hope, held rust briefly in check, then in 1953 and 1954 the disease swept over North America again. By growing one favored cultivar—Marquis, Ceres, Thatcher, or Hope—over so many acres, we made it easy for the fungus to adapt. A cultivar lasted as little as four years before the fungus mutated and outfoxed it.

What is different today is that a new cultivar—a new Ceres or Thatcher or Hope— is much more expensive to develop than it was in 1916 or 1954. The costs are particularly high if the new cultivar is a genetically engineered variety, like Bt corn, which must weather extensive safety tests before it can be planted. A four-year lifespan cannot recoup a company's outlay.

To protect their investment, the seed companies, with the help of the EPA, sought a workable plan to delay—or if possible, to avoid altogether—the onset of resistance to Bt crops. Their solution was one that Shelton had long studied and lobbied for: the establishment of insect refuges or refugia, areas planted in conventional, non-Bt crops, within or adjacent to each cornfield.

In a refuge the pests—the corn borers in this case—are allowed to multiply. Most of them are sensitive, susceptible to the Bt toxin. A larva with the lucky mutation that makes it resistant to Bt enjoys no advantage there in the refuge, because it is never exposed to Bt. It takes its chances along with its kin and might or might not live long enough to metamorphose into a moth, fly away, and mate.

Next door in the Bt fields, however, most insects die before reaching maturity. Among those few that survive and transform into moths, there's a much greater chance that one is resistant—simply because the others have succumbed to the toxin. But with all the normal, sensitive

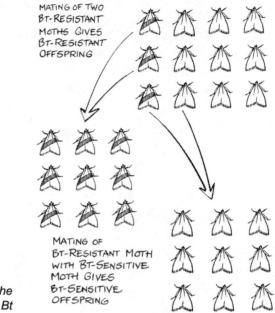

EUROPEAN CORN BORER MOTHS

MATING OF TWO BT-RESISTANT MOTHS GIVES BT-RESISTANT OFFSPRING

MATING OF BT-RESISTANT MOTH WITH BT-SENSITIVE MOTH GIVES BT-SENSITIVE OFFSPRING

How a refuge delays the onset of resistance to Bt

moths now emerging from the nearby refuge, the odds are that this resistant survivor will mate with one of them. The offspring of such a mating should be sensitive to the toxin, having one resistance allele (the gene from the resistant parent) and one sensitivity allele (from the parent from the refuge). When such a heterozygote eats a Bt plant, it dies, eliminating the resistant allele from the population.

It dies, that is, if the resistance trait is recessive, like the wrinkled peas in Mendel's famous experiment. When Mendel crossed peas with round seeds and peas with wrinkled seeds, he found that the seeds of three quarters of the offspring were round. Only if the peas received two alleles for the wrinkled trait would it appear. Roundness was dominant. If the offspring received one roundness allele and one wrinkliness allele, the wrinkliness would seem to disappear, masked by the dominant trait. For the larva of a European corn borer in a Bt cornfield, Mendel's experiment means that a heterozygous insect, one with each type of allele, is doomed because resistance, like wrinkliness, is recessive.

This assumption has been tested by a number of researchers. They selected Bt-resistant insects and allowed them to breed in the laboratory. In almost all of the experiments, the resistance trait was, indeed, recessive. One report in 1999, however, suggested that a kind of resistance that is dominant—which wouldn't be controlled through the use of refuges—can occasionally arise.

To test the refugia strategy under true field conditions, Shelton and his colleagues at Cornell University released diamondback moths into plots of broccoli plants carrying a Bt toxin gene. Shelton chose diamondback moths instead of corn borers for two reasons. First, the diamondback had already shown some resistance to Bt and the trait was known to be recessive. (Because the resistance was apparent by 1993, three years before any Bt crops were on the market, it was clearly a response to the Bt sprays that organic gardeners and others had used for decades.) Second, the diamondback is a warm climate moth. The harsh New York winters would act like the walls of a greenhouse to contain the experiment: any moths that escaped the test plots would die when the cold weather came.

In 1996 and 1997 Shelton and his colleagues monitored the diamondbacks in their Bt broccoli fields. As they reported in 2000, the refuge strategy worked very well—if it was used carefully. In pure stands of Bt broccoli, with no refuge set aside, the moths quickly became highly resistant to Bt. But the larger the refuge, the longer it took for that resistance to develop. It also mattered where the refuge was. A separate, well-defined refuge, around the edges of the field or even up to a half mile away, was more effective in keeping the level of resistance in the moth population low than one created, for example, by planting every other row with the Bt crop.

The separate refuges worked as expected because although diamondbacks travel only short distances in their leaf-eating larval stage, they cover a much wider territory as winged moths looking for mates. When the two types of plants, Bt and non-Bt, were mixed together randomly in a single plot, the diamondback larvae naturally wandered from one plant to the next. The ordinary, sensitive larvae that moved onto a Bt plant and started to munch quickly died. The resistant larvae that found the same plant lived. The result was that only the resistant larvae lived long enough to become moths and breed, passing on their resistance genes. If, on the other hand, a patch of non-Bt broccoli were planted far enough away from the Bt broccoli that the larvae could not wriggle from one to the other, many sensitive larvae also lived to adulthood. Once winged, these sought out and mated with their resistant cousins in the Bt broccoli patch and, following Mendel's laws, the recessive resistance trait was found in many fewer of their offspring. In 2001 the EPA included Shelton's and other scientists' recommendations in its rules for registering Bt corn.

In order to buy Bt corn seed, farmers must sign grower agreements or stewardship agreements, which "impose binding contractual obligations on the grower to comply with the refuge requirements"—that is, farmers who plant Bt corn but fail to establish proper insect refuges can be sued. The onus for educating farmers about proper stewardship—and for ensuring their compliance—falls on the seed company. Any company selling Bt corn seed has to monitor the success of its stewardship plan and report to EPA any "statistically significant and biologically relevant" changes in the corn borer's susceptibility to the Bt toxin its variety expresses. It has to have ready a "remedial action plan" in case resistant insects are detected. And finally, it must submit annual reports to the EPA on its sales, its educational programs, the results of its stewardship plan, and the extent to which its farmers complied with that plan.

In the Corn Belt (but outside cotton-growing regions, where the existence of Bt cotton makes the rules even more exacting), the stewardship agreements between the seed company and the farmer specify a refuge that covers at least 20 percent of the farmer's cornfields (it's 50 percent in the South). This refuge can be laid out as whole fields, as blocks within fields (for instance, as a border along the edges), or as strips, at least four rows wide, across the field. If whole fields, they must be within half a mile (closer is better) of a Bt field.

A refuge can be sprayed for European corn borer and similar pests "only if economic thresholds are reached"; these thresholds are to be determined "using methods recommended by local or regional professionals" such as Cooperative Extension Service agents or crop consultants. Under no circumstances can the spray be any form of Bt. No conventional pesticide has such extensive—and expensive—requirements for managing insect resistance, yet as the case of the diamondback moth shows, insects respond to dusts, powders, and sprays by becoming resistant to the pesticide just as they do when eating Bt corn.

Leaving the refuge unsprayed means farmers must stand by and watch while pests eat 20 percent or more of their cornfields. This kind of compromise, between a short-term sacrifice and a long-term gain, can be hard for a farmer to make. As California grape farmer Victor Davis Hanson writes,

> I once sprayed dimethoate, and in between the stinking loads, as the 500-gallon spray tank refilled, I read the *Inferno,* wondering whether I was in Hell or earning my way. I have dusted organic Dipel six times on vines and watched worms sicken but not die. I have sprayed the toxin Lannate once and watched them drop off before the tractor left the field. Yes, we know it is legal but wrong. Yes, we know it is expedient but nonsustainable for the millennia. Yes, we know that it costs money we don't have. Yes, we know there will be nemeses to confront for doing what we should not; we know that he who pollutes his land must atone tenfold in the hereafter. But farmers as a last resort use terrible chemicals because they do kill, no questions asked. And there are some times in farmers' lives when there are insects, millions of them, that must be killed and killed quickly if their brethren are to eat cheaply and plentifully and on schedule—if one man is to feed 99 other Americans 3,000 miles away by next Tuesday.

In 2000 a telephone survey commissioned by the major seed companies, Aventis, Dow, duPont, Monsanto, and Syngenta, found that 29 percent of the farmers growing Bt corn "broke the rules." The refuges they planted were either too small or too far away. Many farmers pleaded ignorance of the rules. A third of the Bt corn growers in the Midwest and more than half of those in the South couldn't say what size of refuge was required in their area; 60 percent couldn't say how far away the refuge should be; and two-thirds said they didn't know they couldn't spray it with Bt. In 2001 the situation improved somewhat: 13 percent of farmers in the Midwest's Corn Belt and 23 percent of farmers in the South (where the restrictions are tighter because of the proximity of Bt cotton) were "out of compliance." Still, only just over a third, when quizzed, knew the rules for making a refuge in their area. The 2002 growing season saw a closing of the knowledge gap—only 12 percent of the 550 Bt corn farmers surveyed remained unaware of the rules—but 14 percent still didn't comply.

An independent survey by the Center for Science in the Public Interest, found that 19 percent of farms did not comply. Thirteen percent planted no refuge at all. "One reason for the discrepancy," noted the *New York Times*, "was that the industry surveyed only large farms. The center also looked at small farms, which had a higher rate of non-compliance." Those small farms—on the scale of Victor Davis Hanson's fifth-generation family farm—accounted for only 8 percent of the Bt corn grown. Yet, as the Center insisted, the requirement is for each *farm*, not for each county or region.

Before the Center's report came out—and before seeds were sold in 2003—the EPA stiffened the rules. Seed companies are now required to enforce the stewardship agreements the farmers sign by conducting on-farm visits. They are required to help the farmer design an appropriate refuge, and they are required to check, the next growing season, to see if the farmer complied. Farmers who are "significantly out of compliance" for two years will no longer be allowed to buy Bt corn.

In spite of the numbers of farmers who have broken the Bt corn rules, the crop is still winning the chemical battle for survival. Approving the re-registrations of five varieties of Bt corn in 2001, the EPA noted, "Available data indicate that after six years of commercialization, no reported insect resistance has occurred to the Bt toxins expressed either in Bt potato, Bt corn, or Bt cotton products." If the refugia rules are followed, scientists predict that the corn borer will not evolve to be resistant to Bt corn for at least 99 years.

One of Shelton's collaborators, Richard T. Roush, now at the University of California at Davis, believes that the wider use of some techniques already available to plant breeders could keep insect resistance at bay even longer. He suggests designing plants that express Bt toxins only when they are turned on by an otherwise environmentally harmless chemical spray. Or that express Bt only in certain parts of the plant.

But even without these enhancements "the success of Bt crops exceeds expectations," concluded a study published in 2003 by entomologist Bruce Tabashnik of the University of Arizona and seven other scientists, including Shelton. In laboratory and greenhouse tests, three pests have been found that are completely resistant, able to live and reproduce on Bt crops. None of these have yet been found in the field. Even pests living in Bt crop fields that have been continuously monitored for five and six years have shown no signs of an increase in the frequency of resistance.

Discussing the study with a reporter from *Nature Biotechnology News*, Tabashnik said, "If I'd gotten up seven years ago and said that there would be no evidence of increased Bt resistance after Bt crops were planted on 62 million hectares"—meaning the cumulative number of hectares since Bt crops were released in 1996—"I would have been hooted off the stage. No one predicted that there wouldn't even be a minor increase, which is extraordinary."

11   POLLEN HAS
        ALWAYS FLOWN

Labeling gene flow as "contamination," as activists have done, is a misnomer and is a deliberate attempt to provide an emotional tone to a benign natural phenomenon.

—C. S. Prakash (2001)

In October and November 2000, Ignacio Chapela and David Quist collected cobs of *criollo* corn, native varieties of maize grown in the Sierra Norte mountains of Oaxaca, Mexico. The cornfields they sampled, they noted, were "more than 20 kilometers from the main mountain-crossing road." Chapela was an assistant professor in the Department of Environmental Science, Policy, and Management at the University of California, Berkeley. Quist was his graduate student. They collected the corncobs while working at the Mycological Facility in Oaxaca, described by Berkeley's public relations office as "a locally run biological laboratory." Chapela was trained in mycology, the study of fungi. Another press account identified them as advisors for a program that helps indigenous farmers.

Mountainous Oaxaca, along with Chiapas and parts of Guatemala, is recognized as the center of diversity of corn: it was in this southern part of Mexico that teosinte first gave birth to maize thousands of years ago. More different types of corn can be found here than anywhere else on Earth. They are grown in small plots by farmers who "spend a lot of time as breeders," according to Wayne Parrot, a plant genetics professor at the University of Georgia. The farmers seek to maintain and

223

improve the criollo varieties, or landraces. As Masa Iwanaga, at the time the director general of the International Maize and Wheat Research Center in Mexico City (Centro Internacional de Mejoramiento de Maiz y Trigo, or CIMMYT), explains, "The landraces that farmers grow today are often somewhat different from those collected in the same communities decades ago, and they are certainly different from those grown centuries ago, precisely because they have continued to evolve under the combined influence of farmers and the environment. Mexico is not a center of diversity for maize simply because many landraces are 'found' in Mexico. In reality, those landraces are the products of farmers' continuing desire to maintain a great deal of diversity in the maize they grow."

Some of their criollo maize, Chapela and Quist contest, contains the CaMV 35S promoter and a Bt gene from across the border—in spite of the fact that growing genetically modified corn has been banned in Mexico since 1998 expressly to protect the native landraces. In September 2001 Chapela and Quist alerted the Mexican authorities. Government scientists, along with CIMMYT, began their own tests to see if native landraces had been "contaminated" by outlawed plantings. Chapela and Quist published their study in the journal *Nature* in November 2001. A month before the paper came out, the two authors appeared at a press conference with Mexican government officials. The *New York Times* trumpeted, "In a finding that has taken researchers by surprise and alarmed environmentalists, the Mexican government has discovered that some of the country's native corn varieties have been contaminated with genetically engineered DNA."

Rather than sounding surprised, however, one knowledgeable critic said, "It is probably inevitable that eventually engineered genes will be found in Mexican corn, as gene flow is a normal and natural phenomenon with maize." Another said, "I'd be shocked if they didn't find it there." What puzzled the scientists—and inspired activists— were two additional claims in the *Nature* paper: that the DNA was "introgressed" into the landrace, and that it was "attached" to different sequences in different samples, even to different sequences in a single sample. What Quist and Chapela meant by these additional claims, translated a reporter for the *Scientist* magazine, was that the genes "were behaving in a way never before observed: fragmenting into smaller bits of DNA, hopping along the genome like a skipping stone over placid water and potentially creating a tremendous opportunity for mutations. In other words, they claimed that the engineered genes were out of control."

The DNA sequence purported to be so out of control was the CaMV 35S promoter from the cauliflower mosaic virus—a sequence whose job is to turn on the gene next to it. "Activists fears," said a news story in *Science*, "centered on the promoter sequence. . . . If the promoter broke off during hybridization, it could conceivably take over other genes, with unknown consequences." This kind of promoter instability was just what Mae-Wan Ho had predicted in 1999 when she called the use of CaMV 35S a "recipe for disaster." The *Science* story quoted Peter Rosset, co-director of Food First, an advocate organization for small-scale farmers, as saying: "The spread of the promoter could prove to be worse than the spread of the genes for herbicide and insect resistance. If true, this would be a red flag that would call into question every other GM crop on the market."

If true.

Five months later *Nature* took the highly unusual step of announcing that it should never have published the Quist and Chapela paper.

Quist and Chapela had collected ears from four Oaxaca fields and also obtained some corn from Diconsa, a government agency that distributes subsidized food. They pooled the kernels from each ear, ground them to a fine powder, and extracted DNA from the powder. They tested the DNA using PCR to see if it contained three sequences commonly found in genetically modified corn: the CaMV 35S promoter, the NOS terminator from *Agrobacterium tumefasciens*, and the Cry1Ab gene from *Bacillus thuringiensis*. To see the results of the PCR test, the fragments of DNA were sorted by gel electrophoresis. The scientists placed the DNA at one end of a tray of gel and briefly passed an electric current through it. The fragments, pulled toward the positive electrode, sorted themselves by size, because the smaller pieces could travel faster through the gel than the larger ones. The result was a distinctive pattern of bands. Quist and Chapela reported that their tests had detected a short 200-base-pair fragment of the CaMV promoter in several of the Mexican maize samples they tested. On the gels, they had seen a weak PCR band of the right size in four samples from farmers' fields and a strong band in the one obtained from the government. They said they also detected the NOS terminator sequence in two samples and the Bt gene in one.

The PCR bands were weak, they explained, because the transferred gene, or transgene, was present in just a few of the kernels. To a geneticist reading their *Nature* paper, that explanation sounds peculiar. The paper's title was, "Transgenic DNA introgressed into traditional maize landraces in Oaxaca, Mexico." The observation that the transgene was present in only a few kernels doesn't fit with the assertion in the title that the DNA had introgressed. Introgression means that a gene (in this case the CaMV promoter sequence) has not only been introduced into the plant by crossbreeding a landrace with a genetically modified variety, but that the hybrid plants were then repeatedly back-crossed to the landrace. By using the term introgress, Chapela and Quist were claiming that the CaMV promoter sequence was still there, but that the nearby corn genes had been progressively eliminated through the normal crossing-over step that takes place when chromosomes divide. But an introgressed DNA sequence will not be present in only a few of the kernels on a cob; it will appear in all of them.

What other explanations might there be for the weak bands? One possibility is that the DNA samples used to run the PCR tests were contaminated with a bit of CaMV gene. In order to find any bands at all, Quist and Chapela had to use a technique called nested PCR, which entails performing consecutive PCR reactions to detect very small quantities of the target DNA. "This is a particularly risky approach," said the editorial board of the journal *Transgenic Research*, "since extremely low levels of contamination introduced during the handling of samples can be the cause of a positive result."

And Quist and Chapela did, in fact, handle samples that contained the CaMV gene. They used two types of controls to establish the validity of their experiments, to prove that what they saw wouldn't just appear in any randomly chosen corncobs tested using this system. One type, the negative controls, were samples of blue corn from Peru and corn collected by CIMMYT from the same Oaxaca area in 1971. These should—and did—give a negative reading: no sign of the CaMV promoter band. The second set of controls were samples of Monsanto's Yieldgard Bt corn and its Roundup Ready corn. These positive controls should and did give very strong bands in the PCR experiment: a positive reading. Yet the presence of Monsanto's transgenic corn in the same experiment means that there was a chance of contaminating the landrace samples with a minute amount of Monsanto DNA. These contaminated samples would produce a positive reading in the test, but it would be a false positive: a mistake. As several scientists pointed out in letters to *Nature* and, less politely, in the press and over the Internet, Quist and Chapela failed to run the standard follow-up tests to eliminate such false positives.

Quist and Chapela's PCR results are not, however, necessarily wrong. There is still a third possible explanation for the weak bands they saw on the gels—an explanation that does not require introgression or several seasons of back-crossing and yet is not a false positive. The plants from which the cobs were collected could be ordinary landrace plants onto whose silks a few pollen grains from a genetically modified plant growing nearby had landed that same season. Each pollen grain would have germinated on the cornsilk, delivering its sperm cells to fertilize a single egg, which would then develop into the embryo of the kernel. If planted, that kernel would have grown into a new hybrid variety, a cross between a landrace and a transgenic variety. But Quist and Chapela did not plant any of their corn to see if this cross had, in fact, happened. They couldn't because they had ground all their kernels into powder.

Again it was the lack of a double check—not the results themselves—that irritated many scientists. Double-checking would have taken time and patience. More corn would have had to be harvested and planted out the next year. If the gene were present in just a few of the kernels, Quist and Chapela would have had to test many—perhaps hundreds—of the plants grown from such kernels to identify the few plants that were hybrids (if their appearance didn't give them away, which it likely would have). The editors of *Transgenic Research* complained, "Most frustrating is the total failure of the authors to do the easy and incontrovertible experiment of growing out the suspected contaminated lines. Hybrids between Mexican landraces and transgenic commercial maize would be very obvious."

And yet despite the *New York Times'* assertion that researchers had been surprised, Quist and Chapela's critics overwhelmingly agreed that Mexican landraces probably do contain traces of transgenes. Mexican farmers often plant corn kernels—increasingly shipped in from the United States, where genetically modified corn is widely grown—meant as food, not seed. What puzzled the critics most was the claim that the CaMV promoter popped into various places in the genome, that the transgene was "out of control." The *New York Times* quoted a fellow scientist: "If real, that would have been a huge finding."

It was this claim that precipitated *Nature's* announcement that it should not have published the paper. Having received detailed critical letters from reputable scientists, *Nature* asked Chapela and Quist to produce more data in support of their claim. The evidence they supplied, *Nature's* editor wrote in April 2002, "is not sufficent to justify the original paper."

Chapela and Quist's published data were obtained using a technique called inverse PCR, one that a reporter for the *Scientist* identified as "not only complex, but also cutting edge, so even a Ph.D. in another field—no less a layman—would have difficulty in making a reasoned judgment about who is right and who is wrong." In inverse PCR, the primers are inverted. Rather than framing and amplifying the CaMV promoter, they look for what is attached to its two ends. The DNA sample (in this instance from the ground-up corn kernels) would first need to be cut into small bits. As their molecular scissors, scientists use restriction enzymes. When cut fragments are joined back together in a very dilute solution they tend to make circles because the closest matching end to reconnect to is the tail end of the same fragment. When the outward-pointing primers are used to run a PCR test on this solution of circular bits of DNA, what is amplified are the sequences that flank the CaMV promoter sequence they originally detected. These can then be sequenced and identified.

But there are two common problems with the inverse PCR technique. First, the circle-making step is far from perfect, even when the right restriction enzyme is used (and, according to their critics, Quist and Chapela picked the wrong one). No matter how it is cut, the fragment of interest might not find its tail. If it doesn't, there's no telling what other fragment, from which part of the corn genome, it might attach itself to, for the ligase—the molecular glue—is added to a test tube in which thousands of fragments of corn DNA of all different sizes are mixed together. In a solution molecules move. Where a fragment finds itself in the solution bears no relationship to where it was in a corn chromosome. Two pieces that are glued together in solution might have been nowhere near each other originally. This new, double piece might itself form into a circle, making it difficult to figure out that the two fragments were not originally together in the chromosome.

The second problem is that it's quite easy, during the amplification step, to make many copies of pieces that just happen to have some similarity—but are not a perfect match—to the primers. These create what PCR practitioners call artifactual (or bogus) bands. There are ways of making sure a band is not bogus. In Quist and Chapela's case, the piece of DNA on which they based their primer was a 200-base-pair-long fragment of the CaMV promoter. This piece was the one they had detected as weak bands in the first PCR tests on their original corn samples. The whole CaMV promoter is much longer than 200 base pairs. In Bt corn it is more than 1,000 base pairs long. So when Quist and Chapela ran inverse PCR, looking for the bits of DNA just next to the ends of that 200-base-pair fragment, the first thing they should have found was more of the CaMV promoter sequence.

The scientists who published a rebuttal of Quist and Chapela's work simply looked at the authors' own data and asked the question the authors should have asked: are there additional CaMV and other transgene sequences adjacent to the 200-base-pair sequence? There were not. The critics found no similarity to the CaMV promoter except for that very short primer sequence. This result means that the fragments Quist and Chapela amplified in their inverse-PCR experiment were experimental artifacts: mistakes. Quist and Chapela should themselves have suspected this. The critics' analysis was carried out on the authors' own sequence data.

Quist and Chapela claimed that the CaMV promoter broke up into little pieces and got stuck into various places in the corn genome. To people who have created and studied genetically modified plants, this suggestion was quite at odds with experience. However, it would also have been easy to verify. Going back to their original criollo corn DNA samples, Chapela and Quist needed only to use standard PCR with new primers designed to pick up the sequences that the CaMV fragments had supposedly attached themselves to. They did not do this simple experiment. *Nature*'s reviewers should have required this additional confirmation prior to publication. They did not. If the review process had been a bit tougher, it might have saved *Nature* some embarrassment. As for the transgene being "out of control," the *New York Times* reported in April 2002, "Dr. Chapela acknowledged technical problems and said he and Mr. Quist were 'backing off a bit'"—as close as they have come to admitting they were wrong.

But Quist and Chapela stood by their conclusion that criollo corn from Oaxaca, in maize's center of diversity, contained transgenic DNA. Studies by the Mexican government, although not yet published, have also detected transgenes in landraces, as reported at a news conference in Mexico City in February 2002. Yet far from being surprised or alarmed, as the *New York Times*'s original report had it, the researchers and environmentalists associated with CIMMYT have been working since 1995—when the first genetically modified maize reached the market—to understand and contain, if necessary, the effects of these new varieties on Mexican corn. In 1995 CIMMYT held an international workshop called "Gene flow among maize landraces, improved maize varieties, and teosinte: implications for transgenic maize." The proceedings were published in 1997 and are available on CIMMYT's website. CIMMYT has also begun sociological studies to learn how the region's small farmers select seed "and thus influence how genes (including transgenes) flow into and between landraces," said Iwanaga in February 2002.

"In Mexico there is a moratorium on planting transgenic maize," he said. But quite a bit of the corn Mexico imports as food comes from the United States. Since, according to the USDA, a third of U.S. cornfields in 2002 grew transgenic, or genetically modified, corn, Iwanaga said, "It is quite possible that some of the maize imported into Mexico was transgenic." It is also quite possible that some of it is growing in

Mexico today. Mexican farmers are dedicated corn breeders. One could easily have bought U.S. corn, Iwanaga said, "and, instead of eating it, planted it, just to see what might happen."

~~~

Gene flow is popularly considered a hazard of genetically modified food plants. Said Klaus Ammann, curator of the Botanical Garden at the University of Bern in Switzerland, "The debate on genetic engineering 'forces' us to focus in an unfortunate way on gene flow as a basically negative effect, as if pollen would have learned to fly with the transgenes." But gene flow, he said, "has always occurred between different old landraces and between different new varieties of crops. Despite this, varieties of apples or cereals have been stable over many years and specific traits have not disappeared. Pollen has always flown."

Plant genes flow—that is, they move—primarily because pollen is carried by the wind and by insects, sometimes over long distances, to pollinate flowers. Each pollen grain carries two sperm cells, each having a set of the parent plant's chromosomes, ready to fertilize the female. At the heart of each female flower is an embryo sac. The sac consists of a very few cells, all of which are genetically identical. Like the sperm cells, they are also haploid, that is, they contain only one of the two copies of each chromosome in the other cells of the plant. The most important cells in the embryo sac are the egg and the genetically identical "central cell." The central cell gives rise to the endosperm, which nourishes the plant as it germinates and grows until it can produce its own food through photosynthesis. The endosperm, with its stored starch, proteins, and fats, is the primary food of humanity, as well. It's in tortillas and tacos, corn puffs and corn flakes. Wheat endosperm makes flour, bread, and pasta—only wheat germ is the embryos themselves. The animals we raise for meat and a growing fraction of farm-raised fish are nourished by endosperm in their grain- and corn-based feeds.

Pollen—the male contribution to flower sex and what makes people sneeze in spring—is made in huge quantities. A single corn plant makes 18 to 25 million pollen grains. Multiplied by 25,000 or more, for the number of plants in an acre of corn, then by the number of acres in the field, and the amount of pollen released in the short two weeks that corn pollen sheds is truly staggering.

A pollen grain is as small as a grain of sand. Most are round. Some plants make smooth pollen grains, some make patterned ones, and some make spiky pollen. A pollen grain is a vehicle for getting genes from plant to plant. Its surface is designed to resist the drying wind, but to catch and hold tight to the tiny hairs on the female part of the plant. The pollen grain is itself a cell, powered by its own nucleus, called the vegetative nucleus. Inside the pollen grain are two twin sperm cells, each containing the same haploid complement of parental genes. One sperm cell will fertilize the egg. The other will enter the central cell. This process is called double fertilization. It means that both the fertilized egg (the zygote), which will develop into the new plant, and the central cell, which will develop into the endosperm, belong to the next generation.

The landing platform for pollen within the flower is called the style. The style can be a fuzzy central part of a flower or it can be the long silk of the corn plant. Styles are covered with small hair-like projections that trap the pollen carried on the wind and by insects. Pollen is generally quite heavy. Most of it lands within a few yards of the plant that produced it, although some is carried much farther. It lands on everything. Most of it dries up and gets washed off by the rain. Park your car under a tree in spring and it will be coated yellow with DNA-laden pollen cells, dessicated and impotent. Most pollen hits a dead end. Only if the pollen lands in the right place—on the style of a female flower—and only if the chemistry is right, will the pollen have a chance of transmitting its DNA to the next generation.

The chemistry depends on the style. Corn pollen will germinate only on corn silks. Corn pollen that lands on a daisy or on a milkweed plant has the wrong chemistry. The two plants are sexually incompatible. Nothing happens. The pollen dries up and dies. But when a corn silk traps corn pollen—and each long silk traps many pollen grains—then the chemistry is just right. The pollen grain swells and germinates, sending out a pollen tube that grows faster than any other known plant cell—as much as a foot in just a few hours—carrying along the two sperm cells at its tip. The many grains that fall on a single corn silk begin a race down the silk to be the first to reach the egg. Guided by chemical signals inside the silk, the pollen tubes grow in parallel tracks down the silk, aiming steadily for the embryo sac nestled in a little bump on the tiny, unfertilized ear. The pollen grain whose pollen tube grows fastest wins the race, delivering its two sperm cells to the waiting egg and central cell. The next generation begins with the fusion of egg and sperm. The genes of the egg are combined with the genes of the sperm. This familiar process of plant reproduction through pollination is now referred to as "gene flow."

The public's concern about gene flow from genetically modified crops is that the CaMV promoter or a Bt toxin gene or an herbicide-resistance gene might migrate by way of the pollen into wild plants or—in the case of maize—into the landraces. The worry is that these new genes would flow into—newspapers tend to use the words "invade," "contaminate," "pollute," or "taint"—the native corn's genome and, as Chapela said, crowd out landraces that do not carry it.

Yet corn genes have been flowing into—and out of—Mexican landraces for many centuries. Said Masa Iwanaga of CIMMYT, "When transgenes are present in Mexican maize landraces grown by farmers, does this mean that an important resource is lost forever? As scientists, we would answer 'no,' because the landraces may have changed, as they do all the time, but they have not disappeared. On the contrary, with the addition of a transgene, they could actually be considered more diverse. This additional diversity may not be desirable, however. It is precisely this issue that the Mexican government must resolve."

Several months before Quist and Chapela began collecting corncobs in Oaxaca, Juan Pablo Martinez-Soriano of the National Polytechnic Institute in Irapuato, Mexico, wrote a letter to *Science* in which he reminded its readers that corn, unlike its ancestor teosinte, cannot reproduce without human help. Its kernels are tightly fixed to the cob and "viable seeds can only be released by mechanical means (basically by humans). Maize does not disperse itself and therefore does not exist as a free species in nature."

He continued, "Arguments stating that maize is genetically fragile are weak. It seems paradoxical to argue that it is necessary to protect the genetic background of corn when, for 6,000 years of traditional breeding, we have protected only alleles important for humankind. Even if we decide to protect the actual genotypes, there should be no reason for concern. Any transgene transferred inadvertently to native maizes can be removed from the progeny by selecting against the incorporated trait. Maize is always under strong artificial selection, and therefore natural selection has no meaning for the species."

Like corn, all of our food plants originated from wild plants—weeds—with natural means of dispersing their seeds. Some crops and their weedy relatives grow in the same places, some don't. No weedy relatives of corn grow in the United States, Canada, or Europe. But teosinte, the ancestor of maize, grows in Mexico and Central and South America. Teosinte can still make fertile hybrids with maize. When maize and teosinte cross, the next generation is a hodge-podge of traits from the two parents: some plants look more like maize and some look more like teosinte. None is as good a food producer as the carefully bred corn, whether commercial or criollo. But the most important point is that those hybrid plants that receive the maize genes that prevent seed dispersal are sure to die out without humans to pick them, plant them, and tend them.

Long known as cross-hybridization, this problem was only recently dubbed "gene flow." It is not just a problem of crops with weedy relatives. It is an ever-present problem for plant breeders and seed producers. It is most acute in a crop, like corn or canola, that is not a selfer—that does not use its own pollen to pollinate itself—and in crops whose edible parts are produced after pollination, like corn kernels and canola seed. To produce seed for white corn, for example, a farmer has to be sure any neighbors (within pollen-flying range) are not growing yellow corn. The yellow trait is dominant. If white corn and yellow corn are grown side by side, the ears of the white corn will have a mixture of yellow and white kernels. The resulting cross is not always bad. An attractive bicolor variety of sweet corn is sold under several names, including Butter n' Sugar. Yet to keep even that bicolor variety breeding true, it needs to be kept safe from yellow corn pollen.

The same is true for any kind of hybrid corn. Said retired Pioneer Hi-Bred corn breeder Don Duvick, "Of course the problem of unwanted pollination has been around ever since hybrid seed corn production started in the 1930s," he said. "The isolation standards that were set up in those days recognized the fact that zero 'contamination' was a biological impossibility." Corn breeders learned, by trial and error, how far apart two varieties needed to be planted so that crossing would be kept to a minimum.

In order to develop—and sell—pedigreed seeds, breeders had to devise protocols that account for the patterns of pollen dispersal among plants that can crossbreed. Even backyard gardeners had to keep it in mind—and they still do. Anyone hoping to grow the 2000 All-America Selections Winner sweet corn variety named Indian Summer, with its festive mix of yellow, white, red, and purple kernels, needs to be sure to read the fine print: "Requires isolation from other corn pollen."

Whether gene flow between crop plants and their wild relatives is a problem depends on several things: the crop plant, where it is grown, and how it is used; whether or not the crop has weedy relatives nearby with which it can crossbreed; and how those weeds are managed. This long list of issues cannot be lumped—as people have tried to do—into one category. Whether gene flow is a problem also depends on your perspective. Whether you view transgenic corn pollen as a source of genetic pollution or as a source of new genes that could make the corn you're growing more resistant to insects might well depend on whether you're an organic farmer or a subsistence farmer. An environmental activist might see the transfer of genes from crops to wild species as an ecological threat, while a farmer perceives it as a problem in weed management.

What is a weed, after all? One view calls a weed any plant that is growing in the wrong place. It is a plant that is causing harm or being of no benefit. But, notes Klaus Ammann of the Botanical Garden in Bern, it is hard even for botanists to know what plants to list as weeds. "One and the same species may be considered in some parts of its area as a harmless component of natural vegetation, in others as a weed, and again in others even as a useful plant species." Another botanist notes that 17 of the 18 "World's Worst Weeds" are also cultivated. In the right farmer's field, they are not weeds but a crop.

There are technical definitions of weediness, as well. Weeds are plants with characteristics that allow them to persist and spread without human intervention. Weeds have one or more of the following: easily shattered seed pods; seeds that are long-lived, tough, and have structures such as hooks that cause animals to disperse them; toxic chemicals or physical means, such as thorns, of repelling insects and herbivores; or more than one mechanism for reproducing and spreading, such as both seeds and rhizomes. Not surprisingly, crop plants that have been cultivated by people for food over millennia are generally not weedy. They can't readily become weedy because they have lost several weedy characteristics. They are often wholly dependent on humans for survival and propagation. They are, in a very real sense, wards of people. We might worry about milkweed invading the cornfield, but we don't worry about corn taking over the milkweed patch. The lone corn plants towering over the soybean field, called volunteers, present certain weed-control problems, but they're not invasive weeds like Johnson grass, a sorghum relative.

Many crop plants are not native to the places where they are cultivated and their wild relatives are not found in the same geographical areas. The genes of these crops can't escape from the field through pollen, because there are no plants growing nearby that can be fertilized by that pollen. Because there are no wild relatives of corn in the United States, the transgenes the genetically modified varieties contain cannot flow. But that is not true in Mexico, where corn's wild relative teosinte grows. When a crop's weedy relatives live in the same geographical area, often in and near farmers' fields, then the crop's genes can move into the weedy species. Again, whether such gene flow is a problem depends on your occupation and your point of view—as well as on the plant and the trait in question.

Canola, for instance, is defined as the low erucic acid variety of oilseed rape, *Brassica napus*. Canola oil is a health food. Oil of ordinary *Brassica napus* is good for lubricating steam engines, but the erucic acid it contains has been linked to heart disease in people. When he was developing the variety that would become canola, Keith Downey of the Agriculture Canada Research Station in Saskatoon needed to be sure that other brassicas didn't pollinate his breeding plots, introducing the genes that promote erucic acid synthesis. He studied the rate of gene transfer between various types of brassicas, both cultivated and weedy. He found that if a plot of weedy brassica grew within 1,000 feet of a canola field, more than 3 percent of the canola would be fertilized by weedy pollen. The canola pollen itself was not as successful. With as little as 140 feet between plots, only 2 percent of the weeds were fertilized by canola pollen. How far the pollen could fly was not the only factor that had to be taken into account. Was the flower fertilized? Did it produce seed? Did the resulting hybrid grow to reproductive age? Was it fertile? In the late 1980s, Downey did a series of experiments in greenhouses and in the field to answer these questions, as well as to see how easily canola crossbred with its weedy relatives. The various brassicas, whether cultivated or weed, he found, easily crossbred.

Downey discussed his experiments in 1990 at the Workshop on Safeguards for Planned Introduction of Transgenic Oilseed Crucifers held at Cornell University by the USDA. Twenty-three panelists from the United States, the United Kingdom, Canada, Belgium, and India were joined by 70 observers, including representatives from France, Japan, and Thailand. The participants represented universities, industry, the government, and public interest groups. In addition to ecologists and plant breeders, the speakers included experts on herbicides and scientists studying the behavior of honeybees.

The workshop took place three years before the first transgenic crop was approved for market. The participants at the workshop agreed that the question to answer was not "whether or not genes from transgenic species would move out" (they would), but rather when, where, and how it would happen and what would be the consequences.

Because genes that confer resistance to herbicides were among the first to be introduced into crops by molecular techniques, herbicide resistance was one of the first problems associated with gene flow to be brought to the attention of the public. But it isn't a new problem. Herbicide-resistant varieties of canola were already being grown in the 1970s when Downey began his experiments crossbreeding various brassicas. In fact, he used herbicide resistance as his marker to detect crop-weed hybrids. Some of these varieties had gained their resistance to an herbicide through spontaneous or natural mutations; other mutations were induced by conventional plant-breeding methods using chemicals or radiation. Breeders developed them so that canola farmers could have a way to kill the wild brassicas in their canola fields.

When a new herbicide-resistant canola variety was introduced into

Australia in 2000, a study was done to measure gene flow on a very large scale and under real commercial conditions. The variety, which was not genetically engineered, was resistant to the herbicide chlorsulfuron, sold by DuPont as Glean, that inhibits the enzyme acetolactate synthase. The fields surveyed ranged from 25 to 100 hectares in size and were up to several kilometers away from the field planted with the herbicide-resistant variety. The investigators—reproductive ecologist Mary Rieger of the Cooperative Research Center for Australian Weed Management and the University of Adelaide and her colleagues at the universities of Adelaide and of Western Australia—grew the seeds they collected in fields planted with canola that was not herbicide-resistant and asked how many grew into plants that were tolerant of the herbicide. Herbicide-resistant plants were detected in about two-thirds of the tested fields, but their numbers were tiny, averaging 0.03 percent, or just three seeds in 10,000 tested. The herbicide-resistance trait could occasionally be detected at a considerable distance—more than a kilometer—from the field of origin, probably the result of insect-borne pollen.

Several varieties of herbicide-resistant canola are grown in Canada and the United States. Some are resistant to glyphosate (Roundup), others to glufosinate (sold as Rely, Challenge, Finale, or Basta), still others to compounds of the imidazolinone family (Patriot, Lightning, On Duty, and others). Some are genetically modified in the contemporary sense that genes from other plants or from bacteria have been introduced into them by molecular techniques. Others were modified through mutation breeding.

When varieties that are resistant to different herbicides are grown in adjacent fields, they cross-hybridize. Some of their offspring are resistant to both herbicides. This process can continue for generations, leading to what has come to be called "gene stacking." Stacking of herbicide-resistant genes is not a problem for the crop—after all, it's the crop plant that is supposed to survive the herbicide spray. Canola plants have been identified in Canada that can survive being sprayed with three different herbicides. The field in which they were detected had been planted with glufosinate-resistant and imidazolinone-resistant canola in 1997. It was next to a field in which glyphosate-resistant canola was grown. In 1998 plants were detected that had resistance to all three herbicides.

Gene stacking becomes a weed-management problem when the herbicide-tolerance genes accumulate in plants the farmers wish to kill. Because herbicide-tolerant canola was grown in Canada well before the genetically modified varieties were released, farmers already have had to manage herbicide-resistant volunteers that grow as weeds in crop rotations. The most successful strategies employ a long crop rotation cycle, together with plowing and spraying of pre-emergent herbicides.

But weeds related to canola grow in many of the areas where canola is cultivated. In Western Canada, these include *Sinapis arvensis* (wild mustard), *Raphanus raphanistrum* (wild radish), and *Erucastrum gallicum* (dog mustard). There have been no reports from field or greenhouse crosses that these weeds have picked up herbicide-resistant genes from canola. The fact that such gene flow to weeds hasn't been detected might be due to the genetic distance between the crop and its relatives. Experimental crosses between these weedy species and canola produced plants that were sterile or didn't grow well.

In the U.K. as well, despite considerable opportunities for exchanging pollen, the cultivation of oilseed rape has not created hardier weeds. Studies there recently assessed the extent to which oilseed rape (*Brassica napus*) and two wild brassicas, *B. rapa* (wild turnip) and *B. oleracea* (wild cabbage) form hybrids in the wild. None of the wild cabbage plants sampled near oilseed rape fields turned out to be hybrids. Hybrids were detected in wild turnip populations. In the first study one hybrid was detected among the 505 plants sampled; in the second study 47 hybrids were detected among the 3,230 plants sampled.

The low percentage of hybrids detected shows that the *B. napus-B. rapa* hybrids that form every year that oilseed rape is grown have not come to dominate the wild turnip populations. Nor will the introduction of an herbicide-resistant variety of rape affect the wild turnip population. The herbicide-resistance gene is of little use to plants that are not treated with herbicides. Nonetheless, wild turnips are weeds in some rape fields in parts of the U.K. In these fields hybrid weeds will arise. If the hybrids produce seeds, the herbicide-resistant plants that grow from these seeds will be herbicide-resistant volunteers. They will need to be managed by rotating crops and by using different herbicides.

Herbicide use selects for herbicide-resistant weeds just as surely as antibiotic use in humans selects for antibiotic-resistant bacteria. Bob Scott and Chris Tingle are a weed specialist and a soybean agronomist, respectively, for the University of Arkansas's Extension service. They note that 2.5 million acres of soybeans grown in Arkansas contain the gene for resistance to the herbicide glyphosate. "The reason for the large number of acres in Arkansas is the genuine need for the technology," they write. Many weeds commonly found in the state's soybean fields, including Palmer amaranth, common cocklebur, pigweeds, and goosegrass, have developed resistance to one herbicide or another. "Couple that with a long growing season, plenty of rain or irrigation, crop rotational issues, and a large soil bank of weed seeds, and you have a perfect fit for using Roundup Ready soybeans and glyphosate." It is "only a matter of time," they acknowledge, before resistance to Roundup shows up in the weed population as well. When it does, they say, "you need new technology."

Three months before *Nature* abandoned the Quist and Chapela paper, its sister journal *Nature Biotechnology* summarized the Mexican maize scandal. Rather than zeroing in on the technical difficulties, the journal stated that "the major point of divergence in the current discussion" about gene flow was not whether a transgene from a genetically modified variety had moved into the Mexican landraces, but how the so-called contamination would affect biodiversity. The journal wrote: "The Berkeley researchers have claimed that appearance of DNA from GM crops into criollo maize compromises biodiversity. 'If the transgene makes the carrier any more fit,' says Chapela, 'you would expect to see the crowding out of landraces that do not carry the trait.'" A few paragraphs later, the journal gave the opinion of Val Giddings, who is a spokesperson for the industry group called BIO, headquartered in Washington, D.C. Giddings was quoted as saying, "We know what threatens biodiversity, and it is not the substitution of one variety for another in an agricultural field. It is the conversion of native and wild land to agriculture in the first place."

The journal failed to point out that the two sides in this debate were talking at cross purposes. Although each was responding to a question about biodiversity, they defined that term—as their answers make clear—in rather different ways. Chapela is concerned about the diversity of traditional varieties of corn, about the number of genetically different maize landraces planted by farmers in Oaxaca, Mexico. If a genetically modified Bt corn, for example, proved to give a higher yield, it might become more popular than the landraces, and farmers might stop planting these old-fashioned varieties. Whether this change from old to new varieties is bad or good might well depend on your point of view. If the Bt gene were truly introgressed into a landrace, as Quist and Chapela originally suggested, then a landrace would still have all its original genes and it would still grow and produce corn as did the original landrace—but it would give a higher yield because of its new gene for insect resistance.

Yet Chapela's fear—that genetically modified varieties of agricultural crops could push landraces to extinction—is a valid one, though not in the way he is concerned about. According to a United Nations report, the success of commercial varieties of all kinds—not just those created using molecular techniques—has led to the disappearance from farms of more than 80 percent of old-fashioned apples, maize, tomatoes, wheat, and cabbages worldwide. The introduction of the improved varieties has had the positive effect of increasing yields. But the widespread use of just a few different varieties leads to genetic sameness in crops and this genetic uniformity can make them vulnerable to new pathogens or unusual environmental conditions on a very large scale, as evidenced by the wheat rust epidemics.

Preserving crop gene pools is increasingly the function of organizations specifically devoted to the task. In America more than 400,000 crop varieties are maintained in the USDA's National Plant Germplasm System. IRRI in the Philippines has more than 80,000 varieties of rice alone. CIMMYT has collected and preserved hundreds of landraces of maize, periodically growing and harvesting them to preserve their viability. In response to the publicity generated by Quist and Chapela's report, CIMMYT researchers sampled 42 of the landrace populations in its seed bank, finding no evidence that the Bt gene was present in any of the plants tested.

But perhaps this sudden concern with genetic purity is misplaced. In their work with small-scale farmers, CIMMYT's researchers have found that landraces are far from being museum pieces. Population geneticist Julien Berthaud argues that in Oaxaca, where Chapela and Quist gathered their corn, the landraces do not even meet the standard definition of a crop variety. They are not distinct or uniform or stable. Trying to maintain one in a static form could doom it. As a CIMMYT press release explains, "Small-scale farmers select their own seed. Often they choose the best ears at harvest and save seed from only a few cobs—a logical approach but one that increases deleterious mutations. As defects accumulate, the variety loses its genetic value." Rather than preserving a landrace, Berthaud says, what needs to be preserved is "the active flow of genes." If the Bt gene is useful for insect control in landraces, the Mexican farmer who succeeds in introgressing the gene into a landrace will hardly obliterate that landrace but on the contrary will increase its chances of survival.

The idea that transgenic maize will displace landraces is nonsense, says Major Goodman of North Carolina State University. Goodman spoke at a conference titled "Gene flow: what does it mean for biodiversity and centers of origin?" that was sponsored by the Pew Foundation and held in Mexico City in September 2003. Despite the availability of improved corn varieties since the 1930s and the intensive maize breeding that has been conducted in Mexico, Goodman said, there has been little impact on the indigenous landraces grown by 80 percent of Mexican farmers. The new varieties designed for the U.S. Corn Belt are far from well adapted to the subtropics, where they are disease-prone and stress-sensitive. Even though such transgenic varieties have undoubtedly been introduced in Oaxaca, Goodman doubts that it will matter much—they will fare so poorly next to the landraces.

It is the hard struggle for survival faced by the small-scale Mexican farmers who grow and consume the native varieties that is the greatest threat to preservation of the landraces. Said Goodman, "Their economic survival (and hence maize diversity's survival) prospects are bleak, and transgenic maize is probably one of the least of their problems." Adds CIMMYT director general Iwanaga, "The perception that transgenic maize is reducing diversity must not obscure the very real need for research to mitigate the many confirmed threats to maize diversity. Every day, diversity is eroded by habitat destruction, human migration from rural to urban areas, and the irreparable loss of traditional maize seed and knowledge as the farming population ages. The present concern about transgenic maize may only add to these threats. If farmers and consumers are convinced that landraces are 'contaminated' by transgenes and therefore unsafe to grow or eat, farmers will have even fewer incentives to preserve landraces in their fields."

Even if one variety of maize did crowd out another in a Mexican farmer's field, as Chapela fears, it wouldn't have any effect on biodiversity as Val Giddings of BIO was using the term. Giddings was speaking about biodiversity in the much larger sense of the diversity lost when acres of forest are cut down, plowed, planted, and turned into farmers' fields. Rather than thinking about the diversity within a single species—maize—he is thinking about the estimated 10 to 30 million species (not counting bacteria) that inhabit the earth today. In his closing remarks at the Mexico City conference on gene flow, Peter Raven spoke of both meanings of the term. Raven, as director of the Missouri Botanical Garden, is a tireless crusader for biodiversity. He noted that the historical rate of extinction climbed from about 10 per year in 1600 to roughly 100 per year in 1950. It now stands at several thousand extinctions per year.

Biodiversity in Mexico, as elsewhere, says Raven, "can be preserved only in the context of a sustainable nation." Maintaining biodiversity is not as simple as keeping genetically modified maize out of the country. The most serious threats to both maize biodiversity and to biodiversity overall, concludes Raven, are not from GM maize, but from "habitat destruction, urbanization and the abandonment of cultures, alien invasive weeds and pests, and insufficient attention to indigenous peoples and to agriculture in general." In today's world, with its still-growing human population, preserving natural biodiversity requires raising agricultural yields to reduce the demand for new acres to plant. The best modern techniques are needed to increase the nutritional content of crops, while decreasing the impact of agricultural chemicals on wildlife. To Raven, preserving biodiversity also means "increasing the productivity of selected landraces" using contemporary molecular techniques. Far from being a dire threat, the introgression of selected genes that enhance insect and disease resistance might ensure the survival of maize landraces and preserve biodiversity—in the larger sense—by improving small-scale farmers' yields and lives.

12 THE ORGANIC RULE

These mushroom ideas of agriculture are failing;
Mother Earth deprived of her manurial rights is in
revolt; the land is going on strike; the fertility of the
soil is declining.

—Sir Albert Howard (1940)

Sir Albert Howard was a British colonial officer with the cumbersome title of Chemical Botanist to the Government of the Raj at Pusa. Working with poor farmers in India early in the last century, he became convinced that farming was mining the soil of its humus, the dark, partially decayed organic matter that makes soil fertile. In particular, it was using up the nitrogen essential for plants to grow and produce crops. At the Institute of Plant Industry in Indore, India, Howard developed methods for composting animal wastes and urban residue and for using such composts, together with plants grown specifically for the purpose, to improve the soil. He particularly appreciated the importance of mycorrhizal fungi, as well as the fertility-enhancing properties of legumes, plants in the pea and bean family.

Howard believed that the future of civilization depended on the answer to one question: "Can mankind regulate its affairs," he asked in *An Agricultural Testament* in 1940, "so that its chief possession—the fertility of the soil—is preserved?" "We must look at soil fertility," he argued, "as we would study a business where the profit and loss account must be taken along with the balance-sheet, the standing of the concern, and the method of management." He believed that soils deprived of their "manurial rights"—the nutrients not only in animal manure, but also in human wastes—not only lost their ability to support crops and animals, but promoted the spread of disease on the farm. He further advanced the hypothesis that people who consumed produce grown in healthy, well-aerated soils were themselves healthier and more vigorous than people who consumed foods grown in depleted soils, a view echoed by contemporary advocates of what have come to be known as organic foods.

Eve Balfour, one of the first women to graduate from the University of Reading with a degree in agriculture, was inspired by Howard's work. Lady Balfour had bought a farm in Suffolk, England, where she began to compare conventional and "natural" farming methods. In 1943 she published *The Living Soil*; she subsequently organized the Soil Association with other farmers, scientists, and nutritionists, and served as its first president. "The criteria for a sustainable agriculture," she wrote, "can be summed up in one word—permanence, which means adopting techniques that maintain soil fertility indefinitely; that utilise, as far as possible, only renewable resources; that do not grossly pollute the environment; and that foster biological activity within the soil and throughout the cycles of all the involved food chains."

This concept of sustainable agriculture, with an emphasis on natural as opposed to manmade or synthetic means of maintaining soil structure and fertility, has evolved into contemporary organic farming. One of its most important American champions was Jerome Rodale, founder of the Rodale Press. In 1950 he began publishing *Prevention* magazine. Among its missions, he wrote, was "to alert the big-city dweller to the possibilities of obtaining wholesome food either through selective buying or through finding a minute plot on which to grow a few vegetables." He was inspired both by Howard's ideas and by the biodynamic agriculture movement. Biodynamic agriculture grew out of the philosophy of anthropologist Rudolf Steiner, who in 1924 wrote a book called, simply, *Agriculture*. Steiner's approach, like Howard's, focused on soil health, although its composting method relied more heavily on the recycling of animal parts, which today puts it under threat from regulations aimed at reducing the spread of Mad Cow disease.

Another ingredient in contemporary organic farming—the complete avoidance of synthetic pesticides—dates from 1962, when Rachel Carson's *Silent Spring* was published. *Silent Spring* raised awareness about the effects of DDT, a widely used pesticide, on the environment and particularly on songbirds, whose spring voices would fall silent if use of the pesticide continued. *Silent Spring* led to the banning of DDT in this country. For some farmers and consumers, all pesticides were therefore suspect.

The modern organic philosophy expressed by the company Duchy Originals, an organic food retailer founded by Britain's Prince Charles in 1990, is indisputable: "Humans must recognise that humans can only survive and thrive if they live in harmony with the delicate balance of nature between plants, animals, earth, and humans. In farming, that means using methods that conserve and enrich the soil without causing pollution, damaging wildlife, or using up too many of the world's resources. The world's shallow layer of topsoil, on which all our future food depends, contains billions of tiny live organisms in a single handful. These are essential to soil fertility. Protecting the soil and preventing its erosion are essential to our future."

The only point of dispute is how best to achieve those goals.

In its early days organic farming was largely a cottage industry, with consumers buying produce directly from the farmer. This practice continues in some places today. Participants in community supported agriculture groups pay an annual fee for weekly deliveries of organic produce from local farms. But the demand for organic foods grew so rapidly during the last decades of the twentieth century that organic farming became a commercial enterprise, with a clear separation between farmers and retailers. In the 1990s the organic food market was one of the fastest growing sectors of the U.S. food industry, increasing by an average of 23 percent per year. It reached an estimated $7.8 billion by 2000. Organic grocery store chains, such as Whole Foods and Wild Oats, grew and continue to grow, with Solomon Smith Barney predicting a $40 billion market in 2004.

As with any rapidly expanding industry, the central problem for organic food became one of ensuring quality. Given its varied philosophical underpinnings, it had no universal definition of "organic." It was difficult for retailers and consumers to know what they were getting when they bought organic food. Certification organizations had begun to develop in the 1970s, including the Demeter Association and the California Certified Organic Farmers. Many of these were under the auspices of state governments. By 1980 some 40 different organizations were certifying organic farmers and defining standards for organically grown food, each with slightly different rules. Not surprisingly, such differences led to lawsuits.

In 1990, to bring uniformity to this burgeoning business, Congress passed the Organic Food Production Act as part of the Farm Bill. The act established the National Organic Standards Board, with representatives from the food industry, consumers, and environmental groups, and charged it with developing a set of national standards. The bill also established the National Organic Program within the USDA's Agricultural Marketing Service to administer the standards developed by the board. The first National Organic Standards Board was appointed in 1992, the same year that the FDA issued its first guidelines on genetically modified crops. The board issued its first recommendations in 1997.

Popularly known as the Organic Rule, these recommendations took up more than 100 pages in the Federal Register, and provoked an unprecedented volume of comments—several hundred thousand, most of them critical. People objected to the idea that irradiated foods, for instance, could be called organic. The electron-beam technique being used to kill tropical pests on Hawaiian rambutans and other exotic fruits is one form of food irradiation; other forms use gamma rays or X-rays to kill disease-causing bacteria and parasites on spices, wheat flour, potatoes, fruits and vegetables, and meats.

People also objected to the use of synthetic compounds such as antibiotics (if they were not already common practice) in the production of organic foods. And they objected to the use of treated sewage sludge, called biosolids, as fertilizer. This idea had been championed by Howard at the very beginning of the organic movement. It fell under his concept of "manurial rights." In his view, the farm's soil had a right to the nutrients that had been extracted from it when city-dwellers consumed the farm's produce. Fertilizing with night soil—sewage—was necessary, he felt, to keep the soil healthy.

But the one point that produced the loudest protest was the USDA's proposal to allow genetically engineered plants and animals to be used in organic farming. A majority of the National Organic Standards Board had voted to exclude them, yet government policy said that they and their products should be regulated based on risk, not on how they were produced. The USDA proposed that they be treated the same as plants and animals modified by other genetic techniques accepted by the organic community, such as the chromosome-doubling drug colchicine and radiation mutagenesis.

The controversy that followed publication of the draft Organic Rule revealed wide disparities among organic farmers and retailers, insiders and watchdogs. Is organic farming a philosophy of production that makes no claims about food quality? Or is organic food indeed more nutritious, as consumers widely believe? Is organic farming a more sustainable approach to farming than advanced conventional agriculture or isn't it? Is the use of synthetic additives permissible and if so, is the list fixed by very old conventions or can the list evolve in the light of new findings?

Eric Kindberg of the Organic Farmers Marketing Association claimed: "Organic as a market label will be destroyed, both domestically and internationally, if the consumer justifiably reaches the conclusion that an organic product is no healthier than any other labeled product." By contrast, Roger Blobaum of Organic Watch, an advocacy group, objected to USDA's proposal to test organic foods on the grounds that organic farmers had never claimed that their produce was free from all pesticide residues or that they were safer or more nutritious than other produce. Jay Feldman, executive director of the National Coalition Against the Misuse of Pesticides, an organization representing consumers, farmers, environmentalists, and labor, called the USDA draft "a disappointing effort that will have the effect of undermining organic farming practices, environmental protection, and consumer support for the organic label in the marketplace." The notion of organic as a natural way of farming emerged as somewhat at odds with the notion of organic as a source of superior, more nutritious produce, justifying its marketing at a premium price.

In June 1998 the Senate Agriculture Appropriations Committee leaned on the USDA. The committee said it expected the USDA to "construct a National Organic Program that takes into account the needs of small farmers." Fees should be "progressive" (based on income) so that small farmers were not "excessively burdened." Furthermore, the committee ordered the USDA to "follow the recommendations of the National Organic Standards Board, as required by the 1990 farm bill, in issuing final regulations as to what substances are on the national list"—among other things, that meant genetically modified foods had to go.

Secretary of Agriculture Dan Glickman announced that the USDA would revise its Organic Rule in response to public comment and congressional directive. The final Organic Rule was published in 2000. "We listened to consumers and organic farmers and closely followed the recommendations of the National Organic Standards Board to develop a national organic standard that is better than our original proposal," Glickman said. "We believe these new standards fully meet consumer expectations and reflect current organic farming practices."

The final rule specifically prohibits the use of genetically engineered plants and animals, of sewage sludge in fertilizer, and of irradiation in the production of food products labeled organic. It also prohibits the use of antibiotics in organic livestock production and requires 100 percent organic feed for organic livestock. Thus, it was public opinion and the needs of the industry to make a profit that ultimately determined what could and couldn't be labeled organic in the United States, not Howard's or even Rodale's concepts of soil health.

And in spite of the Senate committee's concern that the Organic Rule "takes into account the needs of small farmers," the rule was hardly in effect before small farmers began to see that it did not. "A curious thing happened on the way to a national organic standard: the small farmer, once at the heart of the organic movement, got left behind," wrote Samuel Fromartz in an opinion piece published in the *New York Times* in October 2002. "At local farmers' markets around the country, you'll find many farmers who say their vegetables are 'grown without chemicals' or that their meat is 'free of antibiotics,' but many won't use the 'O' word." Why? "The costs—administrative, monetary, and philosophical—of using the government-defined label are too great." One farmer complained, "After farming for 12 hours a day, I am not going to spend two hours doing paperwork."

Whether grown on small farms or not, whether government certified or not, whether labeled or not, the questions remain: Is organic food more nutritious than food grown with synthetic fertilizers? Does it taste better? Is it safer because synthetic pesticides are not used—or less safe because animal manures are used?

Sir Albert Howard wanted to believe that there was a direct connection between the nutrient content of the soil and the health of humans. That belief persists, if unspoken, among organic farmers and consumers who buy foods bearing the organic label. Although diet is central to health, time and research have consistently failed to establish the link Howard sought. What studies have shown over and over is that diets rich in fruits and vegetables reduce cancer risks—regardless of how those fruits and vegetables are grown.

Neither taste tests nor chemical analyses have been able to consistently distinguish organically grown fruits and vegetables from conventionally grown ones. In Israel in the early 1990s, groups of 40 to 60 consumers were asked to taste fruits and vegetables and to express their preferences. Tasters picked the fresher and riper fruits and vegetables— how they were grown didn't make a difference. The researchers also analyzed the chemical composition of the foods. They found no significant differences between those that were organically grown and those that were grown conventionally.

As for pesticides, a 2002 study commissioned by Consumers Union found that 23 percent of organic fruits and vegetables did contain traces of pesticides, including long-banned chemicals like DDT that persist in the soil. The study included 90,000 samples of 20 crops, a little more than 1 percent of which were organically grown. The scientists did not test for pesticides that are approved for use on organic crops, such as Bt sprays. Of the conventionally grown fruits and vegetables tested, 75 percent showed traces of pesticides.

These results were reported in the press in two ways. An Associated Press story began, "Think organic fruits and vegetables are free of pesticides? Think again." The *New York Times* took the opposite tack: "The first detailed scientific analysis of organic fruits and vegetables, published today, shows that they contain a third as many pesticide residues as conventionally grown foods." The Associated Press said that "the findings don't mean that any of the produce is unsafe. The residues are seldom even close to the limits set by the Environmental Protection Agency." The *New York Times* quoted a spokesman from the American Council on Science and Health as saying that "the amounts of pesticide residues to which we are exposed on our foods pose no significant health risks to human beings." The reporter, food critic Marian Burros, then added, "The Environmental Protection Agency disagrees and has been working to reduce pesticide levels since 1996." Readers of the Associated Press story might begin to question whether organic produce was worth the higher cost, because they are not completely free of pesticides. Readers of the *New York Times* account, on the other hand, would be reassured that organic produce is, as the story's final line states, "a very good way" to reduce your exposure to pesticides.

The point both stories fail to address is, Does it matter? Certainly, eating high concentrations of pesticides makes people sick and can even kill them. So does eating most concentrated chemicals—even aspirin. In the words of Paracelsus, the sixteenth-century Swiss physician considered to be the father of modern toxicology, "The dose makes the poison." The important question is whether the pesticide residues actually present on foods in the supermarket are high enough to cause harm.

Based on risk assessments, the EPA sets what it calls tolerances, maximum permissible amounts for herbicide, fungicide, and pesticide residues in foods. Part of a risk assessment is determining the dose at which adverse effects of a chemical are seen in animals and then calculating the corresponding dose for people. As a report from the Codex

Alimentarius Commission, the international food safety committee convened by the WHO and the FAO, explains, "In most cases ... the substance to be tested is well characterized, of known purity, of no particular nutritional value, and human exposure to it is generally low. It is therefore relatively straightforward," the commission continued, "to identify any potential adverse health effects of importance to humans." The test simply involves feeding guinea pigs or laboratory rats or other animals higher and higher doses of the chemical until the animal gets sick, then setting the human "tolerance" correspondingly lower.

The studies are carried out both by the manufacturers who develop the chemicals and by independent laboratories, and then are evaluated by EPA scientists. The process is far from perfect. There are occasional surprises, both negative and positive. One is the phenomenon called "hormesis." Hormesis is the general name given to the observation that toxic substances—and even other damaging agents such as radiation—have positive health effects at very low doses.

The idea dates to the nineteenth century and was the basis for homeopathic medicine, long viewed with suspicion by the medical community, particularly in the United States. Today hormesis has been indisputably documented, although arguments about its generality continue, as do investigations into its physiological basis. The toxicological implications and the impact on risk assessment are just beginning to be discussed. And there may be important health implications. There is growing evidence, for example, that the increasing incidence of asthma might be the result not of dust and dander in the environment, but of their absence. Small doses of irritants early in life might be necessary to build tolerance. Perhaps if the environment is too clean, the body comes to view normal irritants as foreign and mounts allergic reactions to them.

When pesticides or other chemicals are tested in animals, a common adverse effect of high doses is cancer. "Causes cancer in laboratory rats" is a standard descriptor for a chemical that consumer groups want to see banned. But Bruce Ames, developer of the Ames test for carcinogens, doesn't mince words in summarizing a decade of work on what actually does cause cancer. He says, "The major causes of cancer are: (1) smoking, which accounts for about a third of U.S. cancer and 90 percent of lung cancer; (2) dietary imbalances: lack of sufficient amounts of dietary fruits and vegetables. The quarter of the population eating the fewest fruits and vegetables has double the cancer rate for most types of cancer than the quarter eating the most; (3) chronic infections, mostly in developing countries; and (4) hormonal factors, influenced primarily by lifestyle."

Ames does not believe that we can extrapolate from cancer in rats to cancer in humans. He points out that more than 99 percent of the chemicals people eat are natural. Coffee, for example, contains more than a thousand different chemicals: 28 have been tested, and 19 turned out to be carcinogens in rats and mice. Plants produce many natural pesticides: 71 have been tested, and 37 are carcinogens in rats and mice. Ames further questions the wisdom of trying to reduce what are hypothetical risks of exposure to very low doses of synthetic chemicals. The amount of pesticide residues on and in foods derived from plants, he argues, is insignificant compared to the amount of natural pesticidal compounds. He estimates that Americans eat somewhere between 5,000 and 10,000 natural pesticides, ingesting 1,500 milligrams of such chemicals per person per day—about 10,000 times more than the 0.09 milligram of synthetic pesticide they eat in conventionally grown foods. He concludes, "There is no convincing evidence that synthetic chemical pollutants are important as a cause of human cancer." He states emphatically that "if reducing synthetic pesticides makes fruits and vegetables more expensive, thereby decreasing consumption, then the cancer rate will increase, especially for the poor."

When discussing pesticides on organic produce, Bt sprays are usually left out. Yet the source bacterium of the Bt toxin, *Bacillus thuringiensis,* is a possible allergen. It has also been isolated from burn wounds, suggesting that it might be an opportunistic pathogen. Moreover, *Bacillus anthracis,* which causes anthrax, *Bacillus cereus,* which causes food poisoning, and *Bacillus thuringiensis,* which is the most widely used organic pesticide in the world, all belong to the same species according to the latest studies. The safer way to make use of this bacterium's pesticidal properties is to take the toxin genes out and express them in plants. Plants that produce the toxic proteins, which are harmless to humans, could be used to reduce the use of Bt sprays in organic farming in a way that is completely consistent with the underlying organic philosophy—but for the press of public opinion.

In the debate over the safety of organic versus conventionally grown foods, pesticide residues receive an inordinate amount of attention. More significant food safety issues, on the other hand, are often ignored. Of the factors that make food unsafe, chief in the FDA's eyes today are the microbes that produce toxins. More than 75 million cases of food poisoning are caused by microbial toxins in the United States each year; thousands of people die of it. The microbes responsible include *Salmonella, Listeria,* virulent strains of *E. coli, Campylobacter, Shigella,* and many others. They are found on all raw foods. How that food is washed, stored, and cooked, and whether it is properly salted, pickled, frozen, or otherwise preserved, determines whether or not the microbes will increase to the numbers required to cause food poisoning. The well-known Jack-in-the-Box food poisoning episode resulted in the deaths of three children; 600 people were sickened as a result of eating undercooked hamburgers served by the fast-food restaurant. The meat was contaminated with *E.coli* strain O157:H7. But while we are chary of raw and undercooked meats, American food preferences have been changing to include more fresh fruits and vegetables eaten raw. As consumers have come to prefer buying pre-sorted, pre-washed, and pre-cut salad greens and ingredients in easy-to-use packages, outbreaks of food poisoning have begun to be traced to fruits and vegetables.

We rarely think that salad greens and raw vegetables carry such noxious organisms. They do. Often the contamination can be traced to one supplier. How did it get on the lettuce? The route of transmission is fecal-oral. People's unwashed hands can spread the germs when they handle the food. These bacteria are also present in animal manure. A recent study traced *E. coli* strain O157:H7 to—and into—lettuce from manure through irrigation water. Blocking such routes for infection will receive more attention in the future, both for organically and for conventionally grown fresh foods.

Also important will be finding ways to kill these bacteria on fresh foods, after they are picked and before they are eaten. Of the many means known, some are more effective than others. The best—other than cooking the food thoroughly—is irradiation. If Jack-in-the-Box had used irradiated meat, the restaurant could have served its hamburgers rare with no risk of making anyone sick. Food can be irradiated by gamma rays, X-rays, or electron beams, as is done with the Hawaiian rambutans to clean them of tropical pests.

Both gamma rays, produced by radioactive forms of the elements cobalt or cesium, and X-rays have also been used since the 1950s by plant breeders to induce mutations in seeds. Popular varieties of durum wheat, barley, and rice were created through radiation mutagenesis. The difference between food irradiation and radiation mutagenesis is merely a question of dose. Radiation damages DNA. The higher the dose, the more the damage. If the damage isn't too bad, the cells repair it. Mutations occur because the repair job isn't always done correctly; the repair machinery is error-prone. If the damage is more than the cells can cope with, however, the damaged DNA itself activates a process that causes the cell to self-destruct. Irradiate at a sufficiently high dose, and you kill anything that has DNA in it, including viruses (which are very small targets) and bacterial or fungal spores (which have very tough coats).

If the food contains living cells, as seeds and potatoes do, these are damaged along with the germs. According to a fact sheet from the CDC, "This can be a useful effect. For example, it can be used to prolong the shelf life of potatoes by keeping them from sprouting." For plant breeders, the skill in radiation mutagenesis involves finding a dose that causes mutations but doesn't kill the seed. For purposes of food safety, the right dose is one that kills harmful bacteria but doesn't heat the food enough to change its nutrition or taste.

Food irradiation has been accepted only slowly in the United States, largely because of fears that irradiated food is radioactive. It is not, no more than people become radioactive when the dentist X-rays their teeth. Yet the USDA bowed to public pressure and stipulated in the revised Organic Rule that irradiated food could not be certified as organic. Plant varieties created through radiation mutagenesis, on the other hand, can. And the problem of food poisoning remains.

The eighteenth- and nineteenth-century pioneers of the agricultural sciences—Joseph Priestley, Theodore de Saussure, Justus von Liebig, Friedrich Wöhler, Sir John Bennett Lawes, James Murray, Julius von Sachs—showed that plants were almost magical in their ability to grow on air and water, a few minerals, and sunshine. But pulling nitrogen out of thin air is no small trick. Plants can't do it on their own. They need the help of soil bacteria, some of which capture nitrogen from the air, while others assist in the chemical conversions that free nitrogen from decaying plants and manure.

Among the most remarkable are the soil bacteria belonging to the genus *Rhizobium*. These form partnerships—symbiotic relationships—with peas, beans, and other legumes. Sensing chemicals that the plants' roots produce, they invade the roots, traveling up a thin infection thread into the root's interior. In response, the root forms a small nodule to house the bacteria, lining it with proteins that keep oxygen away from the crucial bacterial enzyme that fixes nitrogen. That enzyme, called nitrogenase, breaks the nitrogen molecule's strong triple bond and converts it to a compound the plant can use. Housing the bacteria costs the plant energy, but the plant benefits because it need not rely on the meager supply of soil nitrogen to make its proteins and nucleic acids. And the soil's capacity to supply nitrogen determines a plant's productivity—how much plant protein can be harvested from an acre of land.

Most of the nitrogen in soil is unavailable to plants. Some is in compounds that plants can't use, some is tied up in partly decayed plants and in manure. Breaking these down, then converting the nitrogen to a form the plants can absorb (a process called mineralization) takes several kinds of bacteria. Both Howard's composting procedures and those of the biodynamic agriculture movement sought to recapture nitrogen from human and animal wastes and, using bacteria, to convert it into a usable form. When these early organic farming principles were being formulated, animal manure was still relatively plentiful. As well, night soil (a polite term for human excrement) was collected and returned to the fields in many places around the world.

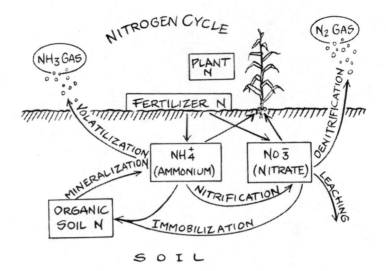

The reactions of nitrogen in the soil

In China it was still being used as fertilizer until the 1980s. But this practice carries the danger of spreading germs and parasites. Most Chinese farmers who used unfermented night soil were infected with parasites.

In his book about nitrogen, *Enriching the Earth,* Vaclav Smil recounts that manure use reached a peak in the Paris suburbs late in the nineteenth century when horse-drawn vehicles provided transportation. Many tons of manure were generated in stables daily and shipped off to farms around cities. For growing vegetables, as much as 500 tons—100,000 pounds—of manure were applied to a single acre of soil. At this extraordinary application rate, even the tiny amounts of nitrogen in manure provided enough to grow luxuriant crops.

With the growth of human populations that followed the industrial revolution, increasing amounts of nitrogen-rich guano (bird excrement) and nitrates were imported by rich European nations from South America to maintain and increase agricultural yields. But it is air that holds a virtually unlimited supply of nitrogen, and many late-nineteenth-century chemists worked on the problem of converting gaseous nitrogen to a form that would support plant growth. Two German scientists succeeded. In 1909 Fritz Haber invented a process to produce ammonia from hydrogen and nitrogen. Over the next few years Carl Bosch developed it into the large-scale process used today. Nitrogen fertilizers made it possible to double and even triple the productivity of land. An official at a sewage authority in South Carolina testified before Congress in 1993 that it would take 50 times the amount of sewage sludge produced in America each year to replace the nitrogen in the chemical fertilizers used in farming.

In Sir Albert Howard's time, land was not a limiting factor. His answer to the nitrogen question, based on his long experience in India, was to grow nitrogen-fixing legumes on any and all spare land. This green manure, plowed under, increased the nitrogen content of the soil, a practice that is a rule of organic farming today. Because synthetic fertilizers are forbidden to organic farmers, productivity is limited by the ability of bacteria to fix nitrogen from the air and their ability to recycle nitrogen from green manure and animal wastes. The nitrogen content of both of these organic fertilizers is very low, and farmers need extra land on which to grow either fodder for their animals or the nitrogen-fixing plants that provide green manure. Because it takes roughly twice as much land to produce the same amount of food, organic farmers must charge a higher price for their produce than conventional farmers do. Countries such as our own and those in Europe enjoy a good climate for growing crops, ample land for their human population, and high income levels. These countries can easily afford to support organic farmers. Such luxuries of land and income are far from universal, however.

People who argue for organic farming—as defined by the USDA's Organic Rule—as a worldwide solution often overlook three points: the higher cost of food, the need for more land, and the need for more hard manual labor.

A six-year study of Washington apples, published in *Nature* in 2001, was said to have "established that organic production of apples is more sustainable than conventional apple production." The researchers, from Washington State University, planted four acres of Golden Delicious apples in 1994 and divided them into plots managed according to organic, conventional, and integrated farming practices. (Integrated farming, they said, mixes parts of organic and conventional methods.) Because yields were equal, the researchers reported that "with the price premiums generally awarded to organic produce, the organic apples were more profitable." The profit margin *without* the price premiums was not discussed.

An outspoken proponent of organic farming in developing countries is Miguel Altieri, a native of Chile and an associate professor of insect biology at the University of California, Berkeley. As an example of how organic farming can work, he cites a 1998 study of organic potato growing in Bolivia. The yield on the organic farms was 11.4 tons per hectare compared to 17.6 tons per hectare on the modern industrialized farms in the same region. The economic gain per ton of potatoes, after the cost of fertilizer and other chemicals was deducted, was slightly higher for the organic farmer than for the high-yield modern farmer. "So organic farming looks pretty good," notes Per Pinstrup-Andersen, director general of the International Food Policy Research Institute from 1992 to 2002.

Until you take a second look. When calculating the yield of potatoes per hectare, the researchers did not take into account the additional hectares of land needed to produce the organic farmers' fertilizer. "Land will have to be set aside either for growing supplementary plants to be used as green manure—in the Bolivian experiment lupines were used—or as acreage for livestock to produce manure." To have enough manure, the organic farmers must either reduce the size of their potato fields or put more land to the plow. "When the cost of the additional land is factored into the study, the figures for yield per hectare do not look so good," says Pinstrup-Andersen. "If we set aside the ecological risks of bringing more land under cultivation, organic farming may be a perfectly acceptable solution in regions with unused land that can be cultivated without damaging the environment." But, he adds, "Such regions are becoming scarce."

In the Bolivian study, 1.5 tons of lupines were used as green manure on every hectare of potato field. The labor involved in growing, harvesting, and applying the lupines to the potato fields was also not accounted for in the cost of producing the potatoes. "In a Kenyan study," Pinstrup-Andersen notes, "for every hectare of land used in maize production, four tons of weeds had to be lugged from hedgerows and roadsides to redress the loss of phosphorous and nitrogen. This is considered women's work." He continues, "Some proponents of organic farming assume that the labor-intensive nature of the farming is itself a good thing." Yet, he says, in parts of Africa, particularly, there is a severe labor shortage, one that is worsening steadily as the AIDS epidemic progresses.

Ebbe Schioler, also with the International Food Policy Research Institute, visited rice growers in Africa. The weeds they faced were "stout thistles, coarse grasses, large thick-leaved plants with tough stalks, and little bushes that . . . produce a powerful, deep-reaching root system." The farmers use no herbicides. "Everything is done by hand and hoe, and even though the children do their bit, it is still touch and go. It takes 40 days of sweating and straining each year to keep just one hectare of land weed-free."

Suggestions that organic farming is appropriate for countries with high population pressures and limited arable land and water supplies sound suspiciously like Marie-Antoinette's "Let them eat cake." Or, as Peter Raven has noted, "Organic agriculture is essentially what is practiced in sub-Saharan Africa today, and half of the people are starving; so it is clear that more is needed."

13   SUSTAINING AGRICULTURE

The true farmer must seek technology and the progress of the ages in order to tame, but not to extinguish, the wild, from which he is nourished even as he exploits it.

—Victor Davis Hanson (2000)

In May 2002 the United Nations Environmental Program released a report on environmental trends. The *New York Times* account stated: "Expansion of cities, destruction of forests, erosion of fields, and rising demand for water are likely to threaten human and ecological health for at least a generation." The growth of agriculture "is damaging landscapes, depleting aquifers, raising the level of salt in the soil, and reducing habitat for wildlife." The *Times* continued, "The report says an important cause is the accelerating growth of vast, poor, and largely unplanned cities in developing countries, most of them near coastlines."

A month earlier, in April 2002, the *New Yorker* magazine published an article called "Leasing the Rain." "The world is running out of fresh water," the cutline read, "and the fight to control it has begun." Between 1950 and 1990, worldwide, the demand for fresh water tripled. By 2025 demand is projected to exceed supply by 56 percent. The fight the *New Yorker* described occurred in "the beautiful old Andean city of Cochabamba, Bolivia," a city of 800,000 people in which "a good part of the population was now in the streets, battling police and soldiers in what people had started calling *la guerra del agua*—the Water War."

263

Neither fresh water nor arable land is inexhaustible, even in the United States. This fact underscores the need for an agriculture that truly conserves both water and land—and *still* grows more food to feed the growing cities. The acreage of land under cultivation worldwide has remained the same for almost half a century. Most of the best agricultural land is already being farmed. On balance, additions—such as acres cut out of the Amazon rainforest for subsistence farms—are cancelled by the loss of prime farmland to urbanization in some places and to desertification and salinization in others. The miracle of the twentieth century was that the amount of food produced worldwide doubled and tripled, while the amount of land farmed stayed the same. This success had three roots: the genetic modification of plants by professional breeders using many different techniques, the increasing use of synthetic fertilizers invented by chemists, and the improved soil and water management methods devised by agricultural scientists and innovative farmers.

The name Green Revolution was coined by William Gaud, the administrator of USAID. "At the time it was an appropriate description of a momentous event," writes Gordon Conway in *The Doubly Green Revolution*. "Today 'Green' signifies the environment; then the image it conveyed was of a world covered with luxuriant and productive crops."

But it is precisely because of the Green Revolution that the world today is not one large farm. Economist Indur Goklany has calculated that if we tried to feed today's six billion people using the mainly organic farming methods of 1961, we would need to cultivate 82 percent of the earth's land surface instead of the current 38 percent. The additional acreage amounts to the entire Amazon Basin, the Sahara Desert, and the Okavango Delta, also known as the Okavango Swamp, an area rich in African wildlife. None of these places is best described as farmland.

Norman Borlaug contends that by improving the productivity of existing farmland, the new crop varieties, fertilizers, and farming techniques of the Green Revolution have saved 20 million square miles of wilderness since 1950. Dennis Avery of the Hudson Institute has pointed out that about 16 million square miles of forest exist today. Forests are the first areas likely to be cultivated when farmland expands (deserts and swamps are not nearly as inviting). "What I'm saying," Avery told the *Atlantic Monthly* in 2003, "is that we have saved every square mile of forest on the planet."

True, the United States currently has a surplus of food, and farmers are paid to take land out of production. But Avery, a former USDA official, has written, "The world has no 'surplus' of farmland, in the U.S., in Western Europe, or anywhere else. The world must virtually triple its farm output in the next 40 years. Inevitably, surplus food stockpiled in America means plowing down more wildlife in some other country." Efforts to protect wildlife and wilderness in other ways will eventually cause conflict. "Nor can we starve the people and let the animals live," Avery writes. "The people would not go quietly. They would not let their children die of starvation while a wildlife preserve sat unplowed next door."

Despite conventional wisdom, the modern contraction of agriculture in the United States has environmental benefits. Gregg Easterbrook, profiling Norman Borlaug in the *Atlantic Monthly* in 1997, labeled the "crisis of 'vanishing farms'" in America as "perhaps the most environmentally favorable development of the modern age." He quoted Paul Waggoner of the Connecticut Agricultural Experiment Station as saying, "From long before Malthus until about 45 years ago each person took more land from nature than his parents did. For the past 45 years people have been taking less land from nature than their parents." And yet, as an article by Jonathan Rauch published in the same magazine six years later pointed out, the percentage of the earth's land surface that is farmed is still rising: "The increase has been gradual, only about 0.3 percent a year; but that still translates into an additional Greece or Nicaragua cultivated or grazed every year."

The Green Revolution has had its ecological downside as well. "Pest and disease outbreaks have been an especially severe consequence," wrote Gordon Conway, due most often to "a combination of factors—higher nutrient levels, narrow genetic stock, uniform continuous planting, and the misuse of pesticides." Erosion and salinization (from too much or improperly designed irrigation) have also been blamed, often legitimately, on the intensive farming practices of the Green Revolution.

Conway continues: "In the 1960s, when the Green Revolution was beginning to make its impact, little thought was given to environmental consequences. They were deemed either insignificant or, at least, capable of being easily redressed at a future date, once the main task of feeding the world was accomplished. There was also a strongly held view, one still commonly voiced, that a healthy, productive agriculture would necessarily benefit the environment. Good agronomy was good environmental management. It is a point with some force," he adds. Farmers do need to be environmentalists.

When farming methods are not ecologically wise, Conway argues, "agriculture is both culprit and victim." In the twenty-first century, the central questions about sustainable agriculture raised by Howard and Rodale and the other pioneers of organic farming remain with us. But are the methods that have come to be called organic—and codified in the Organic Rule—the best we can do?

A hundred years ago when Sir Albert Howard was devising his composting methods, land was not an issue. Between 1870 and 1920 the population increased by 40 percent. The amount of land farmed increased by 75 percent, because more land was cleared, much of it by settlers moving west across North America and by similar expansion in Russia. "The burgeoning populations of those areas were fed by the extension of agriculture rather than by its intensification," T. F. Evans writes in *Feeding the Ten Billion*, "while in Europe there was a marked reduction in the proportion of arable land left fallow."

The human population of the earth was about a billion and a half at the turn of the twentieth century. It is more than six billion today. It is expected to be more than eight billion by 2050. The Great Plains of America and the steppes of Russia are already being farmed. New land could be put under the plow, but arable land is not evenly spread over the globe. More than 90 percent of potential new farmland is in Africa and Latin America. Two countries—Brazil, with 27 percent, and Zaire, with 9 percent—account for more than a third of it. In South Asia, a center of population growth, almost half of the potential farmland is already occupied by cities and towns.

More than half of the people alive today live in cities, with little prospect of growing their own food. By 2020 that proportion will reach

60 percent. They must all be fed. "In 1999 I visited New York City for the first time ever," plant pathologist Jim Cook of Washington State University recalled in 2003. "I took my wife. We had all our lunches at delicatessens. Thousands of people are eating that way. And the next morning all these delis were full again—ham, turkey, chicken, roast beef. It just came home to me: Every night, every city in the world has to refill with food. And none of that food is grown in the city."

He remembered a visit to an agricultural research center near New Delhi, India. "My hosts drove me out of the city. On our way back, it was getting toward six in the evening. Thousands of trucks were parked along the freeway, stacked so high with boxes that they were top heavy. 'These trucks have been moving all day long from the north of India, loaded with perishable foods,' my hosts explained. 'They can't get into New Delhi until midnight, the city is too crowded. They have to wait until folk are off the streets.' At midnight they charge in, unload, then charge out at six in the morning to do it again."

As Wendell Berry says in his 1981 book of essays, *A Gift of Good Land*, "We have an unprecedentedly large urban population that has no land to grow food on, no knowledge of how to grow it, and less and less knowledge of what to do with it after it is grown. That this population can continue to eat through shortage, strike, embargo, riot, depression, war—or any of the other large-scale afflictions that societies have always been heir to and that industrial societies are uniquely vulnerable to—is not a certainty or even a faith; it is a superstition."

Peter Raven uses the concept of the ecological footprint to calculate the impact of today's population on the environment. Dividing the earth's 11.4 billion productive hectares by the current world population—6.3 billion people—means that each person has the use of about 1.8 hectares. According to the Worldwide Fund for Nature's calculations, though, each person in North America uses 9.6 hectares, while each European uses about 5.0 hectares. Only the people in Africa and Asia are living within their means, at 1.3 hectares per person. The worldwide average in 1999, 2.3 hectares per person, is about 22 percent above the planet's carrying capacity. For developing countries to enjoy the American standard of living, Raven adds, "It would take two planets comparable to the planet Earth to support them."

Today discussions of farming methods must take into account the environmental consequences of expanding the food supply further. New land put under cultivation is land taken away from the dwindling wildlands that are the planet's ecological underpinnings—providers of what are called ecological services, such as the underground supply of clean water, climate regulation, and a home for wildlife. If we choose to preserve these, we must ask where the food will come from to feed the still growing human population. If we want everyone to eat as well as today's average American, we must increase our production of food by more than 400 percent.

Thus our challenge for the twenty-first century is to substantially increase our food supply on roughly the same amount of land in production today while simultaneously ameliorating the impacts of intensive farming and putting it on a sustainable basis. Reaching this goal is likely to require improving every aspect of our current agricultural practices—and inventing new ones as well. Yet Ken Cassman, an agronomist at the University of Nebraska, warns that conventional plant breeding and farming methods are already approaching their maximal yields in some places in the world. Moreover, the extreme polarization of current discussions about organic and conventional farming methods stands in the way of using the best of both. We have few unbiased scientific appraisals of competing methods. Most experimental designs, such as the comparisons of Washington apples or Bolivian potatoes, reflect the belief system of the investigators. The notion that valuable insights can be gleaned from both conventional and organic approaches and combined is almost unthinkable—right now— as is using molecular methods to increase yields and protect crops from diseases and pests.

In a book with the uncompromising title *Saving the Planet with Pesticides and Plastic: The Environmental Triumph of High-Yield Farming*, Dennis Avery describes a conference of the Organic Farming Association in 1993.

> I spoke to the conference on biotechnology's potential to produce more food from fewer acres. I pointed out that shifting to organic farming at its current low yields would mean plowing down wildlife habitat equal to the land area of North America. I noted that the world had less than 20 percent of the organic nitrogen needed to support global organic

farming. In response, one hot-eyed organic grower likened biotechnology and farm chemicals to nuclear radiation. Another wanted a "philosophical" decision on biotechnology. I admitted I am no philosopher; I admire biotech purely for its practical ability to save people and wild creatures from famine-related destruction.

Klaus Ammann, curator of the Botanical Garden at the University of Bern in Switzerland, is among the many who believe a compromise could—and must—be reached between organic farming and the new methods in agriculture. In his long career, Ammann has had many "scientific lives," as he puts it. He studied the vegetation of alpine and glacial regions. He mapped the flora of all Switzerland. He studied lichens, specifically their chemistry, and came up with a method of using them to monitor air pollution. He studied urban ecology and weeds, consulted with park systems on how to preserve native flora, and did research in the rainforests of Jamaica.

In the 1990s he founded a group called Ecogene to investigate the biosafety of genetically modified plants. The group completed a million-dollar research project on gene flow in Switzerland in 1996 and published their findings on the Internet. In 2003 he produced a report on biodiversity and agricultural biotechnology for the Botanical Garden at Bern. His conclusion is: "We need organo-transgenic crops. We need to make peace with the people in organic agriculture. We need an ecotechnology revolution."

At the annual meeting of the American Association for the Advancement of Science (AAAS) in 2003, Ammann was asked how scientists could encourage both biotechnology and organic agriculture. He answered, "There's ideology on both sides. In Europe it's just cheap marketing to be GM-free. But to build your marketing on a negative has no future. The next generation of transgenic crops may be more interesting to organic farmers."

Plant pathologist Jim Cook was asked, also at the AAAS meeting, how scientists could encourage biotechnology and organic agriculture at the same time. He replied, "We in the scientific community must never let the divide between the two drive our research." Cook is an adherent of sustainable agriculture, although he admits that even this relatively new term has "taken up all sorts of baggage" and become politicized. To Cook it means an agriculture that is dependent on natural biological cycles, one that places a high value on such resources as soil, water, and energy, on the components of fertilizers, and on genetic diversity, and an agriculture that consciously tries to reduce the pollutants—"the dust, sediments, chemicals, gaseous emissions, and other wastes"—that farming releases into the environment. "These objectives must be met," he stresses, "while continuing growth in the supply of safe, affordable, quality food." When he says "affordable food," Cook has in the back of his mind those cities like New York and New Delhi that, each night, must refill their shops with food.

In 2002 the World Resources Institute published *Fruits of Progress: Growing Sustainable Farming and Food Systems.* It notes that sustainable farming means using "methods that are environmentally sound as well as socially responsible and economically viable. The term 'sustainable' farming may include certified organic practices, and also encompasses other ecological and integrated practices." The publication's case studies cover four wine producers, seven whose main products are vegetables, fruits, and nuts, and one rice grower. All have incorporated basic ecological principles into their farming methods, the institute says, "such as enhancement of diversity (of crops, varieties, soil biota, etc.), recycling and conservation of resources and nutrients, and reduction or elimination of chemical inputs." Cook would not argue with any of these goals; yet none of the farmers in the 12 case studies is attempting to produce high yields of a staple crop at an affordable price and on a sustainable basis, as he is.

Cook's research shows unequivocally that genetically modified crops can contribute to sustainable agriculture. He does not mean virus-resistant papayas or Golden Rice. He means those genetically modified crops most vilified—and least understood—by the press: Roundup Ready soybeans and other herbicide-tolerant crops. He is particularly interested in one that has been developed but not yet marketed: herbicide-tolerant wheat.

For more than 20 years, Cook has been studying no-till agricultural as a way for wheat farmers in the Pacific Northwest to cut erosion and to enrich the soil. Erosion and soil quality are continuing themes in both conventional and organic agriculture. Yet because the Organic Rule forbids the use of herbicides, organic farmers rely on frequent tilling to control weeds. Why do we plow? As Jonathan Rauch writes in "Will Frankenfood Save the Planet?" published in the *Atlantic Monthly* in 2003: "Human beings have been ploughing for so long that we tend to forget why we started doing it in the first place." Before the tractor arrived on the farm, writes T. F. Evans in *Feeding the Ten Billion*, "The guiding adage was: 'When the crop stands still, stir the soil.' Weeds were controlled and soil moisture was thought to be conserved."

Plowing is one of those practices that historian Mauro Ambrosoli has traced, unchanged, through the farming handbooks published between 1350 and 1850—clear back to the poetry of Virgil from before the time of Christ. But research at the Rothamsted Research Institute in England in the 1930s showed that tilling the soil does not conserve soil moisture and that it is unnecessary if weeds can be kept down by "dust mulches" or some other means. "Indeed ploughing, that hallmark of good farming for more than a millennium, could cause substantial soil erosion," Evans writes, "as recognized in the Great Plains of the USA during the 1930s when stubble mulching was found to reduce both wind and water erosion."

Now tillage systems are valued for how much stubble they leave in a field. The best, called conservation tillage, is defined by the Conservation Technology Information Center at Purdue University as "any tillage and planting system that covers more than 30 percent of the soil surface with crop residue, after planting, to reduce soil erosion by water." Achieving this, however, meant putting up with weeds until herbicides were introduced that were effective enough to replace the plow. The first was atrazine, brought out in 1959. Glyphosate, or Roundup, was "the next major step," Evans says, when it was released in 1974, being much less harsh than atrazine. "The reduction in soil erosion by minimum tillage can be striking," Evans writes, "varying from twenty- to a thousand-fold across a range of environments."

Plowing not only dries out the soil and exposes it to erosion, it releases carbon, as carbon dioxide. Carbon dioxide is a greenhouse gas, suspected to cause global warming. "Carbon disappears faster if you stir the soil," Cook explained. "If you chop the crop residue up, bury it, and stir it—which is what we call tillage—there's a burst of biological activity, since you keep making new surface area to be attacked by the decomposers. You're not sequestering carbon anymore, you're basically burning up the whole season's residue." In no-till agriculture, on the other hand, the turnover of organic matter happens in such a way that carbon is sequestered. Said Cook, "You're saving the photosynthate that was manufactured by the plant and returned to the soil as crop residue."

The organic matter in the soil holds more carbon than is in trees or living plants or anything else on land. It influences the amounts of carbon dioxide, methane, and other greenhouse gases in the atmosphere. Plowing releases carbon into the atmosphere; no-till farming keeps it in the soil.

Plowing also, said Cook, "doesn't lead to that nice crumb structure we get when we let the process go slowly." For it is not only the soil's nutrient content that determines its productivity, but its texture or crumb structure, as Cook named it, which is created by the organisms that live in the soil. Cook was talking not only about earthworms (although a study comparing a plowed field with one that had been farmed for 17 years using no-till methods found more than three times as many earthworms in the no-till field), but about fungi and bacteria. Cook explained, "I think what's important is the total biomass, not just who's making up that biomass. You need diversity. You need to have somebody to work in all environments, to have resiliency against stress. The bacteria put out complex carbohydrates that are like glue. The fungi's threads hold the soil together. When they die, they release nitrogen back into the system. There's a lot of symbiosis between plant roots and these microorganisms and we only have the slightest understanding of it."

For instance Dave Weller and Linda Thomashow, Cook's colleagues at Washington State University, recently discovered that a group of microorganisms, different strains of *Pseudomonas fluorescens*, can, as Cook put it, "team up with roots to protect them against disease—it's most interesting. We used to just think the nitrogen-fixing bacteria and the fungal mycorrhizae were the good actors. Here we have another whole world that's providing benefit to the plant."

When Cook first became interested in no-till in 1974, it was not a popular way to farm, even though it had been invented by what Cook calls progressive farmers. It is also known as direct seeding, because the seed and fertilizer are placed in the soil at the same time. Notably, the field is not plowed first. Planting involves one pass with a tractor towing a seeder with two tools. Said Cook, "One disk makes a slice, a zone an inch or two wide and three to four inches deep. The second disk is just a little shallower." The first slice is for liquid fertilizer, with the proportions of nitrogen, phosphorous, and sulfur determined by a soil test. The second disc places the seed.

Direct seeding, Cook said, encourages the growth of soil microorganisms that decompose matter on the surface, "where there is a much more dynamic environment of alternating light and darkness, ultraviolet radiation, and wetting and drying." When a field has made the transition, as Cook put it, from a plowed field to a direct-seeded field, "the seed-drill goes through the soil easier, and the straw rots faster on the soil surface. The straw that's left behind with one harvest disappears by the next spring." It is what he calls "composting in place." An acre that he has direct-seeded for more than 20 years "is organically as beautiful as it can be. The soil is mellowing out with better crumb structure. You can pull the plow through it in third gear." In spite of the beauty of the soil in his test plot—and the savings in labor, time, and gas from having to ride a tractor less—Cook finds it has not been easy to persuade either his colleagues or the region's wheat farmers that no-till is the future of farming. "Farmers are reluctant to change, and so are scientists," he noted.

The Conservation Technology Information Center reports that in 1991, when Iowa farmers were asked why they didn't switch to conservation tillage to control runoff and erosion, they answered: weeds. Other surveys of farmers have had the same result. If they knew they could control weeds without plowing, farmers would readily convert to no-till.

The biggest increase in no-till farming since 1996 has come in soybean farming. The center believes it is no coincidence that Monsanto's Roundup Ready soybeans, genetically engineered to tolerate the herbicide glyphosate, were introduced in that year. In 2001 the American Soybean Association randomly surveyed soybean farmers in 19 states who planted 200 or more acres. They found that the number of acres being farmed with conservation tillage methods, including no-till, had jumped from 25 to 83 percent. Compared to 1996, more than half the farmers plowed less, while 73 percent left more crop residue in their fields. Why such a change? Sixty-three percent of the soybean growers said, without being prompted, that the reason was Roundup Ready soybeans. They could control weeds in their fields without plowing. In 2002 75 percent of the soybeans planted in American were genetically engineered herbicide-tolerant varieties, the majority of which were Roundup Ready.

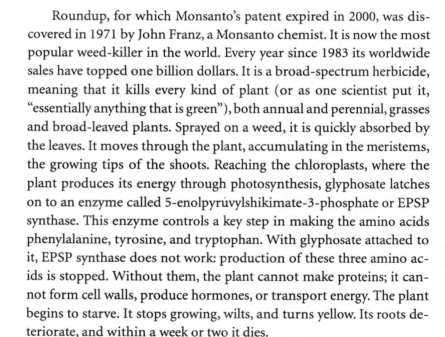

Roundup, for which Monsanto's patent expired in 2000, was discovered in 1971 by John Franz, a Monsanto chemist. It is now the most popular weed-killer in the world. Every year since 1983 its worldwide sales have topped one billion dollars. It is a broad-spectrum herbicide, meaning that it kills every kind of plant (or as one scientist put it, "essentially anything that is green"), both annual and perennial, grasses and broad-leaved plants. Sprayed on a weed, it is quickly absorbed by the leaves. It moves through the plant, accumulating in the meristems, the growing tips of the shoots. Reaching the chloroplasts, where the plant produces its energy through photosynthesis, glyphosate latches on to an enzyme called 5-enolpyruvylshikimate-3-phosphate or EPSP synthase. This enzyme controls a key step in making the amino acids phenylalanine, tyrosine, and tryptophan. With glyphosate attached to it, EPSP synthase does not work: production of these three amino acids is stopped. Without them, the plant cannot make proteins; it cannot form cell walls, produce hormones, or transport energy. The plant begins to starve. It stops growing, wilts, and turns yellow. Its roots deteriorate, and within a week or two it dies.

Roundup does not harm insects, fish, birds, or mammals (including humans) because none of these creatures have the enzyme EPSP synthase. Unlike plants, animals do not make these three amino acids (phenylalanine, tyrosine, or tryptophan) but instead take them from their food. Fungi and bacteria, including the ones in the soil to which Cook attributes good crumb structure, do contain EPSP synthase. Yet soil organisms can—and rapidly do—metabolize glyphosate, breaking it down into carbon dioxide and ammonia.

In 1994 *Farm Chemicals* magazine included Roundup in the "Top Ten Products That Changed the Face of Agriculture" because, they said, Roundup encouraged conservation tillage. Jim Cook agrees. He testified before Congress in 1999, "I can say unequivocally that the development of Roundup as a tool has been the single greatest tool for moving forward to growing crops with less tillage." Yet it wasn't a perfect tool to begin with. Glyphosate has no way to distinguish one plant from another, a crop from a weed. So while it could clean a farmer's fields of weeds before planting, it couldn't be used while the crop was in the ground. Other herbicides were more specific. Grasses like corn and wheat can tolerate 2,4-D, which mimics the action of the auxin growth hormones. It kills only broad-leaved plants like dandelions. In a soybean field, trifluralin (which inhibits cell division in the roots) can be used; corn sprayed with it, on the other hand, won't produce a single ear.

Could Roundup be made more specific, made to target only some plants and not all of them? Monsanto's Ernie Jaworski began working on the idea in the 1970s. "I was just curious about why this chemical killed all plants but didn't injure animals," he said when asked how he had gotten the idea for Roundup Ready crops. Steve Padgette, an enzymologist, was hired by Monsanto in 1984. "Jaworski had a clear vision that resistance to glyphosate would be valuable," he recalled. "He just knew if we could do it, it would work."

By 1984 Monsanto researcher Ganesh Kishore had identified a change in the enzyme EPSP synthase that would make a plant tolerate Roundup. Said Padgette, "I walked in and started working on petunia EPSP synthase. We were getting some pretty good tolerance but not commercial levels of tolerance. The 'aha' moment, for me at least, came in 1986 or '87. We had a big collection of bacteria that would degrade glyphosate. So we got the idea to screen the bacteria for EPSP synthase. We used a robot. We found one particular bacterial culture in which the extracts showed really good results. It had super high efficiency."

By the end of 1988 the team had isolated the gene and created a vector with which to introduce it into soybeans. "We were working all through Christmas getting these vectors ready to go out," said Padgette. The results were good: "We had soybean transformation." This gene, from the familiar *Agrobacterium*, is now found in all Roundup Ready crops: alfalfa, canola, cotton, corn, lettuce, potato, soybean, strawberry, sugar beet, sugarcane, and wheat.

Besides Monsanto's glyphosate-tolerant varieties, other crops, known by the name Liberty, have been developed that can tolerate being sprayed with a different herbicide, a compound called glufosinate, sold as Basta, Rely, Challenge, and Finale. Like Roundup, glufosinate is a broad spectrum herbicide, killing all green plants, but its method of action is quite different. Glufosinate was first discovered in a soil organism called a streptomycete, which produces it to poison competing soil microorganisms. Sprayed on a plant, it interrupts the production of the amino acid glutamine, which the plant makes from the nitrates in fertilizers, both natural and synthetic. The nitrates are first turned into ammonia, from which an enzyme called glutamine synthase makes the amino acid. Glufosinate disables this enzyme. The plant cells continue making ammonia until it reaches toxic concentrations. At the same time, having too little glutamine, they can no longer fix carbon. The combination quickly kills the plant.

Because humans and other animals do not use ammonia to make glutamine, glufosinate is not harmful to them. The streptomycete, however, does use ammonia; it has developed a way to avoid poisoning itself. The streptomycete produces an enzyme that attaches an acetyl group onto glufosinate, chemically inactivating it. It can no longer bind to glutamine synthase and so cannot disrupt the production of the amino acid. To make glufosinate-tolerant crops, the gene that allows the streptomycete to inactivate the herbicide was cloned and transferred, using the *Agrobacterium* method, to crop plants. Canola, chicory, corn, cotton, rice, soybean, sugar beet, tomato, and wheat have all been modified to be glufosinate-tolerant.

A third kind of herbicide-tolerant crop can withstand spraying by imidazolinone herbicides, sold as Patriot, Lightning, On Duty, and other brands. Imidazolinones also interrupt the production of amino acids, in this case leucine, isoleucine, and valine. The enzyme blocked is acetolactate synthase (ALS). When the herbicide binds to ALS, the enzyme stops working, so the amino acids can't be made. The plant, unable to produce proteins, dies. The ALS enzyme can readily mutate to a form that is not sensitive to imidazolinone. It takes a change of a single amino acid in the protein's sequence. For this reason, imidazolinone-tolerant crops can be created without genetic engineering. Those currently on the market, such as corn, canola, rice, and wheat, were created by chemical mutagenesis and somaclonal variation.

Having three kinds of herbicide-tolerant varieties of wheat available to farmers, Cook believes, could give the same boost to conservation tillage in the Pacific Northwest that Roundup Ready soybeans did in the Midwest. Yet so far two of these varieties have stayed on the shelf, the companies choosing not to commercialize them because of political opposition to genetically modified foods. "In Washington, we're a big wheat-producing state," Cook explained, "and 90 percent of our wheat is exported. It's a different kind of wheat than is grown in the Midwest. It's low protein, only 8 to 9 percent protein. It works well in chapatis and noodles." Wheat growers have resisted switching to no-till farming using genetically modified crops because, said Cook, "That would make our wheat a GMO, a genetically modified organism, and so far unacceptable to our international customers."

Wheat that is tolerant of imidazolinone, however, is not genetically modified. "BASF used mutagenesis, old-fashioned mutagenesis," Cook said. To mutate the gene responsible for the ALS enzyme, BASF exposed seeds to the chemical sodium azide. "It beats up the DNA," Cook explained. "Some seeds will be dead, others will produce plants with all kinds of morphological changes, and if you don't like them, you throw them away." The next step is to spray the healthy-looking seedlings with the herbicide, throw away the ones that die, and continue growing the rest. "You compare them back to your unmodified source," said Cook. "You look and screen and throw away, and then scale up. You don't know how many genes have been changed besides the one targeted to make the plant tolerate the herbicide. You don't have a clue. You might have changed a hundred, or many hundreds of genes. And unless a change can be recognized by looking at the plant, or by watching its performance in the field, it will go undetected." An imidazolinone-tolerant wheat variety called Above, in the line known as Clearfield, was released to seed producers by the Colorado and Texas experiment stations in 2001 and by the Oregon experiment station in 2002. Having been conventionally bred and not modified by molecular techniques, these varieties are not subject to federal regulations or international conventions concerning genetically modified foods.

Said Cook, "I'm delighted to see Clearfield come in. I chuckle under my breath to hear that mutagenesis is considered safe and genetic engineering is not, but if that's what society gives us, we'll take it." In the Pacific Northwest, where Cook is located, farmers grow both spring wheat and winter wheat. He can foresee a three-year, no-till rotation using Clearfield winter wheat, Roundup-Ready spring wheat, and Liberty spring barley. Each cereal is herbicide-tolerant, but each has a completely different mode of action to the other two. "It would buy all kinds of durability into the system," he said. But, he warned, "You need to be careful with Clearfield. That ALS gene is easily mutated." Already some weeds are resistant. "We'll see tolerance in four to five years, but if we can then use Roundup Ready or Liberty crops, we can extend its usefulness."

Cook concluded, "The principle we need to never forget is that diversity is just as important in agriculture as in the environment. This three-year combination would be so efficient you could just skip the herbicide altogether one year—the ground would be so clean."

Ecologists have taught us that the natural environment provides many essential services: purification of the water and of air, mitigation of droughts and floods, and protection of biodiversity. Asked Cook, "Can't we do these same things while farming the land? The answer is yes, with no-till. You improve soil structures, stop erosion, sequester carbon, improve water filtration, rather than letting it run off the land, and store more water in years of drought." The stubble left behind provides habitat for birds and small mammals, which could lead to an upsurge in the number of their predators, including hawks, owls, and coyotes. Compared to conventionally tilled farm land, Cook considers no-till "a whole new ecology" and "a huge step toward being environmentally benign and toward contributing services with social value." Farmers can achieve these goals, he believes, with a three-year rotation of herbicide-tolerant cereals using three different herbicides— Clearfield, Roundup, and Liberty. But they can't do it with just one.

14　SHARING THE FRUITS

I believe that scientists, as a privileged group of citizens, have more than an academic responsibility to advance science. They must also accept a higher social responsibility and, wherever possible, use science to help solve the important problems not of industry, but of humanity.

—Ingo Potrykus (2001)

Writing in the *New Yorker* magazine in April 2000, Michael Specter tells the story of the San Marzano tomato. It's a plum tomato, from outside Naples, Italy, where the soil is sweetened by the ash of Mount Vesuvius. "According to Neapolitan tradition, pizza was invented as a vehicle for the consumption of the San Marzano," Specter writes. Yet farmers of this prized tomato are losing the battle against a virus, the tomato fern leaf virus, better known as the cucumber mosaic virus or CMV. It "has taken over the fields," farmer Eduardo Angelo Ruggiero told Specter. "The tomato was born here. Now I think it's dying here. We understand that genetics could help, but the question is political. I myself have mixed feelings. I am afraid that if we grow tomatoes differently they will taste like every other tomato in the world. But there is also a truth. We have lost 90 percent of our production in the past decade." In 20 years the Naples region dropped from being the number 1 tomato producer in Italy in 1980 to number 4 or 5. Ruggiero himself has stopped growing tomatoes. His children, who had planned to continue the family business, "will have to do it somewhere else," Specter writes.

In Adams County, Pennsylvania, in March 2000, farmer Jim Lerew of Lerew Brothers Orchard had to bulldoze and burn all 150 acres of his peach and nectarine trees. His neighbor, Jim Lott of Bonnie Brae Fruit Farm had to destroy 25,000 trees—227 acres of orchard. Lott's trees would have produced $400,000-worth of fruit that year, he estimated. In each case the reason was plum pox virus. The tree-burning was the USDA's attempt to contain the attack: Lerew was the first farmer in America to identify plum pox as the reason his peach trees looked sick.

Known as sharka (from the Bulgarian for "pox"), plum pox is well known—and hated—in Europe. Some hundred million European peach, nectarine, apricot, plum, cherry, and almond trees are infected. Sharka was found in Chile, also a major fruit producer, in 1992 and is now widespread in that country. Its symptoms resemble those of a nutritional deficiency or insect damage. Often only a few leaves on an infected tree show signs of sickness, yet the fruit is discolored, misshapen, sour, and small. Eventually a sick tree stops bearing fruit altogether. In an infected orchard, crop losses can reach 100 percent. For Pennsylvania that could mean the loss of a $22.5 million industry. If the disease were to spread throughout the United States, losses could climb to $1.8 billion.

For Lerew and Lott the problem is more personal. It could mean losing their family farms. According to Lerew, "Financial losses are higher than a year of lost production. It takes six years for a new fruit tree to start producing in volume and it costs a lot to remove and replace the tree itself." The government paid Pennsylvania fruit growers $1,000 an acre to destroy their orchards. For Bonnie Brae Farms that was $227,000 for trees that each year produce $400,000 worth of fruit.

Throughout Pennsylvania more than 900 acres of orchards were bulldozed down and burned in the spring of 2000. For the farmers, deciding what to do next is difficult. Replant with peaches or switch to plums, and they risk a recurrence of the pox. The disease is incurable, and no cultivars are yet resistant. Apples are a choice—the pox does not affect them—but many of these farmers already have some acres in apples. They had planted peaches in the first place to lengthen their harvest season, and to give more job security to the migrant workers who pick the crops. Picking apples is a three-month job. Add in peaches and plums, and the job grows to seven or eight months long. Other crops are not suitable for these hillside farms, nor do the farmers have the capital to buy the equipment needed to plant and harvest something new. The most attractive option to the owner of a burned orchard might well be the ready cash to be had by selling out to a housing developer.

Pennsylvania has invested $5.1 million in its efforts to eradicate plum pox virus. The USDA has allocated nearly as much. Part of the money is supporting research on how to control the pest, which has two known vectors, two ways of traveling from place to place: humans and aphids. "Undoubtedly the human vectors are much more effective in spreading the disease," says plant pathologist Fred Gildow of Penn State. "Legally imported fruit tree cultivars are tested for disease at USDA quarantine nurseries before being released for public use," Gildow remarks. The carriers of plum pox virus, therefore, most likely came in illegally, "by accident or out of ignorance," by someone who failed to inform Customs that they had visited a farm or carried a twig in their luggage.

Like the cucumber mosaic virus, the plum pox virus is naturally carried from plant to plant by aphids. An aphid picks up the virus while tasting a plant with its stylet, a syringe-like mouth part. If the sap tastes right, the aphid settles down to feed. If not, it retracts its stylet—now contaminated with the virus—and flies off to taste another plant, which then becomes infected. The virus itself is a single molecule of RNA wrapped in a protective protein coat. Injected into a plant cell, the RNA takes over the cell's protein-making machinery and instructs it to make virus particles instead.

In Europe scientists studying the plum pox life cycle have counted 3.5 million aphids of 40 different species in each acre of orchard; 20 species of aphid are known to carry plum pox virus. Each tree is visited by 50,000 to 300,000 aphids every year. It takes only one aphid to infect—and kill—a tree.

Insecticides cannot kill every aphid in an orchard. Nor are the harmful species necessarily those that colonize peach and plum trees. In France and Spain the culprit is usually a "migrant aphid," one that

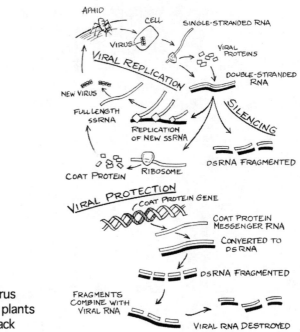

How a plant virus
spreads, and how plants
resist their attack

does not usually live in peach trees and is simply passing through the
orchard looking for a suitable host. The bird cherry-oat aphid, for example, prefers to feed on small grains; when the grains are harvested,
the aphid migrates and can be found in large numbers in peach orchards in late summer. In France the bird cherry-oat aphid carries plum
pox virus. Whether it does in Pennsylvania has not yet been tested. Nor
is it easy for a farmer to tell just when the bird cherry-oat aphid might
be passing through and it is time to spray. Aphids are tiny and look
much alike. Besides, as a USDA scientist noted, insecticides don't kill
on contact: the migrant aphid can still infect a tree before it dies.

"The key is to stop them before they take flight, or try to disrupt
their life cycle," says Vernon Damsteegt, a plant pathologist at the
USDA's Foreign Disease-Weed Laboratory in Maryland. One way is to
keep orchards completely weed-free, denying the aphids a place to overwinter. To be effective, the weed-free zone would have to be larger than
an aphid usually flies after tasting (and rejecting) a plant. According to

Gildow, 100 yards is generally the maximum distance a virus will be carried from an infected plant to a new host. A buffer zone of 300 yards should keep an orchard pox-free. Keeping farms weed-free, however, has its own drawbacks.

In 2003, the *Philosophical Transactions of the Royal Society* in Britain published the results of what it called the Farm Scale Evaluations and the press called the GM crop trials. The trials included 66 fields of sugar beets, 68 cornfields, and 67 canola fields, each planted half with a genetically modified herbicide-tolerant variety and half with a conventional crop that was not herbicide tolerant. Teams of scientists from three research centers counted and compared the number of species of weeds and of weed seeds, as well as of beetles, butterflies, and other invertebrates found inside and along the edges of each half-field throughout the year. They concluded that some genetically modified crops (canola and sugar beet) reduced biodiversity, while GM corn increased it.

The difference was that the chemicals currently used to keep weeds out of conventional cornfields (atrazine and its derivatives) are quite effective—better at killing weeds than the glufosinate used with the genetically modified corn. The herbicides used with conventional canola and sugar beet, on the other hand, are less effective than glufosinate when used with GM canola or than glyphosate when used with GM sugar beets. The studies thus, as the *Scientist* reported, "are less about GM crops directly than about the herbicides used to manage weeds in GM and conventional varieties." GM corn was said to increase biodiversity because the herbicide used on GM corn (glufosinate) did not kill as many weeds as the herbicide used on non-GM corn (atrazine).

Herbicide-tolerant crops can be created through conventional breeding. Herbicide-tolerant varieties of canola, corn, and wheat created through chemical mutagenesis and somaclonal variation are currently sold. Whether the crop is genetically modified by mutagenesis or by molecular techniques, the crop-herbicide package that is most effective will result in the fewest weeds—which is, indeed, what farmers are hoping for when they buy weed-killer. But the fewer the weeds, these studies found, the fewer the butterflies, beetles, and other invertebrates living in and around the field—except for those that feed on dead plant matter.

Because 70 percent of the land area of the United Kingdom is either planted in crops or grazed by farm animals, it's understandable that the British scientists and press should define biodiversity as the number of butterflies, beetles, and other invertebrates that live on the weeds growing in and alongside farmers' fields. Yet such a definition puts the Pennsylvania peach farmer in a quandary. Aphids are among those other invertebrates that the British believe farmers should protect as part of desirable biodiversity. Aphids carry the plum pox virus that can wipe out a Pennsylvania orchard and leave the farmer with little alternative but to quit farming and sell out. A housing development or a shopping center would most likely be lower in biodiversity than an orchard, even one enclosed by a 300-yard weed-free zone.

Yet there is a way to have both farmland biodiversity—weeds and their invertebrate residents—and orchards free of plum pox. The answer resides in three plum trees growing under quarantine in a USDA orchard, three trees genetically modified to be resistant to the plum pox virus.

Named C-5, the resistant plum tree was developed by Ralph Scorza, a USDA horticulturalist known for breeding the Bluebyrd plum. C-5 contains a gene, provided by French researcher Michel Ravelonandro, that codes for the plum pox virus's coat protein. Coupled with a strong promoter, the coat protein gene provides the plant with the equivalent of an immune response—although it works very differently from the immune response in humans.

Inside the cells of the resistant plum variety, the coat protein DNA is transcribed into messenger RNA. So much messenger RNA is made, in fact, that it activates an oversupply shutdown mechanism. The mRNA is destroyed and the coat protein is never made. When the plum is infected by the pox virus, the virus uncoats its RNA, ready to instruct the cell to make more viral RNA and more coat protein to wrap it in. But the cell is ready. It recognizes the RNA and destroys it. The virus-resistant papaya, which in 1998 saved the Hawaiian papaya industry, is protected by this same post-transcriptional gene silencing mechanism. Dennis Gonsalves, who developed the resistant papaya, provided Scorza with the *Agrobacterium* vector used to transform the plum.

The C-5 plum passed the new gene—and the resistance—to its offspring, so it could be used to breed other plums by crossbreeding or grafting. In fact, when cuttings were grafted onto a rootstock infected with the plum pox virus, C-5 depressed the virus so much that aphids could no longer pick it up and spread it from tree to tree. The rootstock was cured. As of September 2001 the C-5 plum had been tested in three European countries for five years and it was still resistant. But it remains "under development" and has not been commercialized. Once it is on the market, the USDA's magazine *Agricultural Research* wrote, plum farmers "can breathe a sigh of relief, knowing they have a backup tree to plant. And U.S. consumers may prefer a transformed plum to no plum at all."

Relief perhaps very much like what Hawaiian papaya growers felt when they learned that American consumers would, indeed, prefer a transformed Hawaiian papaya to no Hawaiian papaya at all. Three years after the virus-resistant papaya seeds were first given out, Hawaiian papaya production—and sales—were back up to what they had been in 1994, before the virus moved into the prime papaya-growing region of Puna. The 2001 harvest of more than 55 million pounds was worth $14 million to Hawaiian farmers.

The industry recovered in spite of what became known in the media as The Great Papaya Scandal, or "How Prof. Joe Cummins uncovered the great scandal of how U.S. regulatory agencies approved a GM papaya even though it carries a viral gene known to be a potential allergen." Cummins, a retired professor, had previously been a coauthor with Mae-Wan Ho of the 1999 opinion piece, "Cauliflower Mosaic Viral Promoter—A Recipe For Disaster." He read a scientific paper that reported a slight similarity between the papaya ringspot virus coat protein and a protein known to be allergenic. Apparently without understanding that post-transcriptional gene silencing *prevents* the coat protein from being made, he became alarmed that papayas might now be full of allergy-causing proteins. He contacted an official at the FDA, who told him the resistant papaya was the regulatory responsibility of the EPA. He learned that the EPA had granted the papaya an exemption from regulation—no doubt because a virus-resistant papaya contains much less viral coat protein than the virus-infected papayas people had been eating. Cummins, still unaware of the facts, began his crusade. If he had understood why the plants were protected from the virus, he wouldn't have had a story.

In 2002 Dennis Gonsalves and his research team were awarded the Alexander von Humboldt Award for making "the most significant contribution to American agriculture" in the previous five years. In 2003— the year Hawaii's genetically modified papayas were approved for sale in Canada—Gonsalves won the Leadership in Science Public Service Award from the American Society of Plant Biologists. The citation pointed out that Gonsalves was born and raised on a sugar plantation in Hawaii. "What inspired our team," Gonsalves said, "was the knowledge that we had to apply the best science we could to solve very real problems for farmers and families who were desperate for a solution."

The Italian farmers growing the prized San Marzano tomatoes are also desperate for a solution. But for many of them the only solution possible is to stop growing San Marzanos. Francesco Sala of the University of Milan reported in April 2003 that "already today, the San Marzano accounts for just 3 percent of Campania tomatoes, while 20 years ago this figure was 30 percent. Today a lot of growers use an American hybrid in its place which yields well and has a good taste, but is not the same thing. In short, the San Marzano is a product in extinction."

Italian researchers, back in 1993, used the same basic method with which Gonsalves developed Hawaii's new papaya, and Scorza Pennsylvania's new plum, to modify the San Marzano tomato to be resistant to the tomato fern leaf virus. But whether a transformed San Marzano tomato is preferable to no San Marzano tomato at all was still to be learned 10 years later because, says Leonardo Vingiani, the director of the Italian biotechnology association, Assobiotec, the "necessary authorisation" for field trials had not yet been granted.

Assobiotec's Sergio Dompé writes in the magazine *Slow: The International Herald of Taste*, published by the Slow Food movement, "Could the name be the problem?" The words "genetic engineering" and "biotechnology," he argues, call up "a glaring contradiction between life and technology, the natural and the artificial, that generates concern and apprehension." He compares acceptance of genetically modified foods to the acceptance of nuclear magnetic resonance—the medical procedure that suddenly became noncontroversial when its name was changed to magnetic resonance imaging. "The moral of the story: Inappropriate words, such as a misunderstood adjective or a bold juxtaposition, often influence our view of reality, feeding our suspicions and unspoken fears even when there is no justification."

Italians, he says, fear that biotechnology will standardize or "dumb down" their traditional foods and flavors, jeopardizing the country's heritage. "This is a very serious accusation," Dompé says. It is also, he continues, "as unjust as it is serious. It is a myth."

In Italy, he writes, rare plant diseases can be devastating. Neither conventional farming practices nor organic methods, for example, have been able to save the San Marzano plum tomato—which Dompé calls "a national treasure on a par with Piedmontese wines"—from the attack of the cucumber mosaic virus. "In this case," Dompé says, "genetic engineering has offered more than just the promise of a solution. It has already proved able to save the crop by inducing virus-resistant characteristics. The fact that it is not yet possible to apply this solution in the field (because it is based on 'genetic modification'), suggests that sometimes the risks for typical Italian products do not derive from biotechnology but from unreasonable resistance to their use. It would indeed be a tragedy to condemn this tomato variety to extinction, simply because of preconceived ideas."

~

The loss of the San Marzano tomato will be a tragedy to those who prize its distinctive flavor. Pizza might never be the same. If lack of the San Marzano drives small family farms like Eduardo Angelo Ruggiero's out of business—as a virus threatened to do in Hawaii and still threatens in Pennsylvania—more people will be affected. Some will miss eating fresh locally grown produce. Farm families must find another income. Environmentalists who hope to keep the land open and undeveloped will lose another bulwark against the shopping center and housing complex.

For Florence Wambugu the issues surrounding the virus-resistant sweet potato are in some ways the same—she hopes to keep Kenyan women producing a good crop of sweet potatoes—and in other ways very different. In Kenya the failure of the sweet potato crop does not mean just the lack of a prized taste or even a monetary setback for otherwise well-off and well-fed farmers. In Kenya the word tragedy retakes its true meaning. Lacking sweet potatoes, a family can starve. Interviewed by the *New Scientist* magazine in 2000, Wambugu explained, "The sweet potato is a major staple crop. It is always there in the backyard if there is nothing else to eat."

Sweet potato supplies more food energy and more nutrients than any other crop that could be cultivated in the same amount of (mostly marginal) land. Yet the yield per acre in Africa is one third that in China and less than half the global average. In 1999 Matin Qaim of the University of Bonn authored a study of the potential economic effects of genetically modified virus-resistant sweet potatoes in Kenya. Sweet potato, he reported, is cultivated on 75,000 hectares in Kenya, mostly in plots of less than half an acre with two crops a year. Forty percent is kept for the farm family to eat. The other 60 percent is sold, harvested a little at a time, "whenever cash is needed to meet the basic household requirements." It is considered a woman's crop. Although men might help with the plowing, the weeding and harvesting are left to the women—as are any sales, either to traders who come to the farm periodically or by carrying baskets on their heads to the nearest village. If yield could be increased, those women and their children would have more food and more money.

A genetically modified virus-resistant sweet potato, Qaim predicted, could increase yields by 18 percent. (Wambugu believes that yields could double.) If yields increased 18 percent, farm incomes would rise by 28 percent, Qaim calculated, for a total of $5.4 million across the country. "Because women often control the revenues from sweet potato sales," he wrote, "there is a high probability that a significant proportion of the additional income would be spent on food."

The difference could be critical to a hungry child. According to political scientist Robert Paarlberg, 30 percent of the children in Africa are chronically malnourished due to "lagging productivity on small farms." In Africa, he says, more than anywhere else in the world, the problem of hunger is due to farming methods that simply do not produce enough food. In some parts of Africa, he writes in *The Politics of Precaution: Genetically Modified Crops in Developing Countries*, "yields are actually declining from their already low levels." Although farmers cut down nearly 5 million hectares of forest a year to expand their fields and pastures, the amount of food produced per person has not increased since 1970. Even counting food aid and commercial imports, each person has less food to eat now than in 1980.

Africa's farm production problem is not the result of modern technology. The agricultural advances of the past half century have largely passed Africa by. As Florence Wambugu says, Africa "missed" the Green Revolution. It can't afford to miss the next one. Asked by an interviewer for the *New Scientist*, "Surely what African farmers really need is fertilisers and better irrigation? Won't putting money into GM technology divert attention from these more basic needs?", Wambugu answered: "I think that is like saying Africans don't need aircraft, we should go by road. Or that we should be denied computers until everybody has bought a typewriter and mastered it. We are part of a global community. Of course, we need to look at why existing agricultural technologies have had so little impact in Africa. Africa needs to pick and choose technologies, to learn which ones are compatible."

Wambugu's own choice of technologies is influenced by both her upbringing and her education. "As one in a family of nine children growing up on a small farm in Kenya's highlands," she says, telling her story in the *Washington Post*, "I learned firsthand about the enormous challenge of breaking the cycle of poverty and hunger in rural Africa. In fact, the reason I became a plant scientist was to help farmers like my mother, who sold the only cow our family owned to pay for my secondary education. This was a sacrifice in more ways than one because I, like most children in Kenya, was needed on the farm."

It was, as she puts it, "to solve a national problem" that Wambugu joined the Kenya Agricultural Research Institute upon receiving her degree in botany from the University of Nairobi. There she worked with scientists from the International Potato Center on the problem of viruses in sweet potatoes. To learn more, she traveled to North Dakota State University for a master's degree in plant pathology. She earned her Ph.D. in virology and biotechnology from the University of Bath in England in 1991, doing the fieldwork for her dissertation on sweet potato viruses at home in Kenya.

Using traditional plant breeding methods, she tried to improve the sweet potato's resistance to viruses and pests. "But after years of hard work and frustration," as she puts it, "I finally realized I would not be able to develop a virus-resistant potato through traditional plant breeding." It was then that she was invited to visit Monsanto.

Roger Beachy, then at Washington University, had collaborated with researchers at Monsanto in the late 1980s to produce the first virus-resistant plant by inserting a virus coat protein gene. In 2003 Beachy and Monsanto's Robert Fraley and Stephen Rogers received a patent on this technique, which is the one used to create virus-resistant papaya, plum, and San Marzano tomato, as well as squash, sweet potato, cassava, and several other crops. Rob Horsch, currently Vice President for Product and Technology Cooperation at Monsanto, recalled, "It was Ernie Jaworski that got the idea that it looks like this stuff—both biotechnology broadly and virus coat protein gene specifically—would work. He thought it ought to be shared." Jaworski's studies of how Roundup worked had led Monsanto into the business of plant breeding in the early 1980s.

Horsch continued, "Ernie talked with Joel Cohen at the U.S. Agency for International Development (USAID) and hatched a plan to do some kind of root or tuber virus protection project." They chose roots or tubers because these vegetables are susceptible to the most severe virus problems, Horsch said, "and because no other technology was solving the problem." Monsanto wanted to recruit an African scientist to decide which root or tuber, and which virus, the project should focus on. Said Horsch, "We figured an African expert would be better qualified to know what was most needed in Africa, rather than making that choice in St. Louis. As we began the recruiting process, Florence Wambugu's name popped up."

Wambugu arrived at Monsanto headquarters in 1992, bringing

seven sweet potato varieties from Kenya. She joined Horsch's lab at Monsanto and worked for two years constructing the vector—the ring of DNA containing the virus coat protein gene, the promoter, and the accessory DNA needed for it to be accepted by a sweet potato plant cell—that would make the plant resistant to the sweet potato feathery mottle virus. When her time was up she returned to Nairobi; since then a series of visiting Kenyan scientists, Horsch says, have been "steadily making progress on this difficult problem."

Monsanto donated to the Kenya Agricultural Research Institute (KARI) not only the improved sweet potato plants but also the license to use for free whatever patented genes or techniques were needed to create them. KARI has since sub-licensed the technology back to the Donald Danforth Plant Science Center in St. Louis. The center, whose director is now Roger Beachy, has become the Kenyans' primary technical partner in place of Monsanto.

In 2000 Horsch visited the first field trials of genetically engineered sweet potatoes in Kenya. "The first field test revealed what virtually all first tests reveal," he has said. "More work is still needed to produce a satisfactory product. The commercial biotech products my company began selling in 1996," he explained, "are the result of going back to the drawing board four or five times."

According to Lawrence Kent of the Danforth Center, the field trials didn't do well because the virus gene came from an American strain of the virus, not a local Kenyan strain. "Even Monsanto knew there was a good chance it would fail, but they wanted to test the system," he explained. After the disappointing performance of the first gene, Monsanto created three new vectors and donated them to both KARI and the Danforth Center. The new vectors use a virus isolated in Kenya combined with a new technique, called codon optimization, that increases the level of expression of the coat protein gene. "With sweet potato, the problem is mostly scientific at this point," Kent said. Only testing will show if the new gene gives sufficient protection against the virus.

Thirteen years after Florence Wambugu arrived at Monsanto headquarters, a virus-resistant sweet potato is still not growing in Kenyan farmers' fields. "The project has not moved rapidly," a USAID spokesperson told *Science,* "because KARI doesn't have the expertise to manage transgenic crop development."

Robert Paarlberg provides a gloss on that term *expertise.* "It is extremely difficult for politically cautious leaders in poor countries to be seen welcoming GM seeds if they are coming from a private corporate lab in the United States," he says. "In fact, a strong pattern emerges." In researching his book, *The Politics of Precaution,* he learned that one reason Kenya has not approved the virus-resistant sweet potato is that the technology came from Monsanto. One reason Brazil hasn't approved Roundup Ready soybeans and India hasn't approved Bt cotton is that they are Monsanto products. And one reason China did approve Monsanto's Bt cotton was that it could at the same time approve varieties made by Chinese scientists.

For Horsch, who is now in charge of a project called Growing Partnerships at Monsanto, "It also illustrates the dilemma between doing it and trying to build up the local capacity." By capacity, he means both training and access to the necessary tools and techniques. KARI's plant biotechnology laboratories, according to a Danforth Center scientist, are now first-rate. "They have established transformation capacity at KARI that's equal to ours," said Nigel Taylor. "They're doing very well now. Really it's a collaboration now, not a technology transfer." And yet the decision to help Florence Wambugu and her fellow Kenyan scientists learn how to create virus-resistant plants has, said Horsch, "come at a price." He explained, "The unfortunate consequence is the product is still not done. If Monsanto had decided to do it ourselves, the product would have been done three to four years ago. Sweet potato was both the wrong crop—the Rockefeller Foundation picked rice to fund—and too experimental to attract much development assistance money from public sources."

Yet Horsch also alludes to Paarlberg's insight when he concedes that the project was finally transferred completely to the Danforth Center because, "There was the need for the program to stand on its own merits, and not be seen as a Monsanto project."

Agricultural economist Lawrence Kent, a former Peace Corps volunteer with 10 years' experience working in Africa, was hired by the Danforth Center to help it increase its collaborations with African scientists. "Living in Africa," he said, "I saw that better seeds and other improved agricultural technologies could make a big impact on people's lives." He was working in Egypt when he first became aware of the Danforth Center. "I saw they had good science and a good mission, but they were missing a step: transferring their technology to Africa. In Africa you have to really work the process. I proposed that they hire me to help, and Roger agreed."

Among Kent's most difficult tasks is getting approval for a field trial. One problem is that the regulatory system is not in place. "Lots of countries are working on it, but Kenya is about the only country in sub-Saharan Africa, other than South Africa, that has an application form for a field trial ready for you to fill out. Even in Kenya, there are no permanent bodies to give approvals. They have to convene a committee each time. They all hear these stories from Europe. They have these half-issues and doubts in their minds."

Wambugu gave her opinion in an essay published by the *Washington Post* in 2001: "So the question becomes, why aren't these types of biotechnology applications more readily available to African farmers? I believe blame lies with critics who claim that Africa has no chance to benefit from biotechnology and that our people will be exploited by multinationals. These critics, who have never experienced hunger and death on the scale we sadly witness in Africa, are content to keep Africans dependent on food aid from industrialized nations while mass starvation occurs."

Asked by *New Scientist* magazine in 2001 if she was "an apostle of Monsanto in Africa," she replied, "Some people say I am fighting for the company. But I say I am a stakeholder in the technology. It is 20 years of my life. I believe in the benefits it has for our people. So I fight for the credibility of the technology."

Two years later, when she continued the discussion with an interviewer for the Canadian Broadcasting Corporation, virus-resistant sweet potatoes were still not on the market. "GM sweet potato has nothing to do with Monsanto as a product," she told the television audience. "It is not of any value to any company that I know of and I think, if anything, it has served a really good example on how African countries can help themselves by partnering and going back and developing their own system so they can make decisions."

The interviewer asked, "Dr. Wambugu, you're trying to persuade us all that genetically modified foods are the way to go for agriculture in Africa, right?" Wambugu answered, "I don't think that my business is to persuade anyone specifically on GM. What we're looking for in Africa is all kinds of technologies that can help us to increase food production. And GM is only one of these technologies."

15 FOOD FOR THOUGHT

You people in the developed world are certainly free
to debate the merits of genetically modified foods,
but can we please eat first?

—Florence Wambugu (2003)

Visitors to Monsanto headquarters in St. Louis in 2003 stopped on their tour to admire a potted YieldgardPlus corn plant, a variety that can fend off both the corn borer and the corn rootworm. According to Eric Sachs, Monsanto's director of Scientific Affairs, "These are the only two corn pests that need pesticide applications." Each is a billion dollar pest for American corn farmers. "Both problems were solved by Monsanto," said Sachs.

Asked why he considered YieldgardPlus such a breakthrough, Sachs explained that traditional methods—even crop rotation—no longer hold the pests in check: "What has emerged over the past five years is that the insects have developed ways to get around crop rotation. Instead of laying their eggs in a cornfield, they lay their eggs in a soybean field, which is the rotation crop. So the larvae hatch in a cornfield. The second way is to lay eggs that overwinter more than one year. They don't hatch in the soybean field, they hatch in the cornfield after the rotation." Eventually, he agreed, the insects will become resistant to the Bt toxin produced by YieldgardPlus. "We're already working on the second generation plant, producing a second Bt protein that targets the same insect. It makes it much more difficult for the insect to become resistant. It needs to have two rare mutations at the same time."

YieldgardPlus, like earlier versions of Bt corn, will benefit farmers and make money for the company—not trivial accomplishments. But it isn't likely to improve the image of genetically modified foods—or of Monsanto.

A very different variety of corn, casually mentioned by Monsanto vice president Rob Horsch, however, inspired the *St. Louis Post-Dispatch* to print an editorial, "Genetically Modified Crops Feed the World," in which the new corn is called a boon to the world and an example of the technology's greatest promise. Horsch calls it Golden Corn. It has pale kernels with brilliant orange embryos, half moons at the kernel's heart. "It's a white corn with a golden embryo," he explained. "It's really quite beautiful." Like Golden Rice, it has been modified to produce more beta carotene, the precursor to vitamin A. White corn was chosen as the starting material, because it is preferred in Africa, where vitamin A deficiency is a problem and corn is a staple crop. "It has an embryo-specific promoter," Horsch said. "But there's enough packed in the embryo that the whole kernel has higher beta carotene than yellow maize."

According to Horsch, Monsanto has said it will donate the rights to Golden Corn to the African Agricultural Technology Foundation (AATF), a nongovernmental organization developed with funding from the Rockefeller Foundation. "Farmers will be able to use the seed without paying Monsanto," the *St. Louis Post-Dispatch* explained.

Initially thought of as a patent bank, the AATF was designed, according to its brochure, "to resolve many of the barriers that have prevented smallholder farmers in Africa from gaining access to existing agricultural technologies that could help relieve food insecurity and alleviate poverty." It describes itself as "the neutral intermediary, a 'responsible' party between owners of proprietary technologies and those that need them."

Florence Wambugu was a member of AATF's Design Advisory Committee, along with representatives from seven African countries. USAID was represented, as were its counterparts in Denmark and the U.K. Gerard Barry represented Monsanto on the committee; he left the company late in 2003 to become head of the Golden Rice project at IRRI in the Philippines.

Yet while Monsanto, Dow AgroSciences, Pioneer Hi-Bred, and Aventis CropScience helped set up the foundation, Horsch points out that "the AATF board of directors doesn't include any of the companies." The companies donate licenses for patented genes or technologies; AATF owns any product that results. "For Golden Corn," said Horsch, "we've donated the basic set of genes to enhance beta carotene." Associates at Ohio State University and at CIMMYT, the international maize and wheat research center in Mexico, will use the genes to transform corn plants, analyze the results, breed the trait into popular varieties, and see them through the regulatory process.

Horsch enthusiastically identified this coalition of public and private entities as "we." "We want to combine beta carotene with the high protein maize already developed," he said of the future. "The problem with the high protein trait is that you can't see it, you can't taste it." But combined with the high beta-carotene trait, the protein trait becomes visible: the plants with orange embryos will also be high in proteins.

Golden Corn is just one example of a new trend in plant breeding, called biofortification. The idea was popularized by economist Howarth Bouis of the International Food Policy Research Institute, one of the 16 Future Harvest Centers under the World Bank's Consultative Group on International Agricultural Research. "The idea is to breed plants for higher nutrition content," said John Beard, a nutritionist at Penn State University. In May 2003 Beard had attended a meeting in Cali, Colombia, of some 70 plant breeders, nutritionists, economists, and community activists. Beard reported: "Ten years ago Bouis had the idea that supplement and fortification programs, from an economics aspect, need constant investment to get a response. He wondered, Is there some way to frontload the system? To make a big investment at first, and then tail off? Why not breed staple crops for higher nutrition?" Each of the international plant breeding centers allied with the Food Policy Research Institute was asked to start screening their varieties for micronutrients, not simply for hardiness or high yield. Said Beard, "You want to shift a whole population to a different plane of nutrition. You won't see benefits right away because the content of the micronutrient is usually quite small; hence, the improvements will occur in small increments over a long time."

Karel Schubert at the Donald Danforth Plant Science Center, for example, is working with nutritionists at Tufts University and the University of Florida to fortify rice with folate. Folate (or vitamin B-9) and iron are the most important micronutrients whose deficiency leads to anemia, impaired cognitive development, and neural tube defects such as spina bifida, Schubert explained. "In Asia there are 100,000 cases of neural tube defects a year. In China it's called 'the disease of the winter marriage,' because the mother is deficient in folate prior to conception. It's also the most important birth defect in the U.S." Spinach is a prime source of folate. Legumes and some fruits are high in it, but all cereals, including rice, said Schubert, are "seriously low in folate." So are root and tuber crops, like potatoes. However, unlike the beta-carotene pathway, how plants make folate is not yet well understood. Schubert and his colleagues are studying the folate pathway in the common laboratory plant *Arabidopsis*; they have much to learn before any folate-rich crops will be ready to market.

Plant breeders are increasingly paired with nutritionists in such biofortification projects because simply raising the amount of folate or other nutrients in a plant isn't enough to enrich it: the nutrient must also be in the right form. A bioavailability study is needed to measure how easily the human body can take up the nutrient and make use of it. Penn State's John Beard, along with Jere Haas at Cornell University, became involved in the biofortification of crops after plant breeders at IRRI discovered an iron-rich variety of rice.

The rice, called IR68144, had been developed using conventional breeding techniques to grow well in poor soils and cold temperatures. When Bouis first announced his biofortification challenge, IRRI plant geneticist Glenn Gregario began screening the 80,000 varieties of rice in the IRRI germ-plasm banks. He grew 2,000 different varieties, harvesting the grain and analyzing it for iron content, before he discovered IR68144 in 1998. Because it wasn't altered by molecular techniques, IR68144 was not subject to any safety testing. It was grown in quantity immediately and used in feeding trials. It advanced through cell-culture tests and a pilot study of 27 people to a full feeding trial involving 300 nuns in a convent in Manila. If not for the typhoons that twice destroyed the harvest, proof of the rice's nutritional advantage (if any) would have been available in 2000.

As it was, by 2003 Beard and his collaborators at Cornell University had analyzed 10 different parameters of nutrient status on each of 1,080 blood samples from the 300 nuns. They measured levels of vitamin A, iron, folate, vitamin B12, and zinc. The blood tests were combined with direct dietary analyses. "We weighed everything the sisters consumed three times randomly every two weeks throughout the course of the nine-month feeding trial," Beard explained. "Their meals were analyzed for micronutrient content by using the Philippine Food Tables, which tells us that so many grams of this food contains so much iron, zinc, etc. Then, based on the amount of vitamin C, coffee, and tea consumed with the meal, we can calculate the bioavailability of the rice. Then we ask, What happens when we switch the forms of rice in the diet? Does the iron status change?"

According to an IRRI publication, "The trial of IR68144 is being widely regarded as an attempt to prove the concept that staple foods enriched with micronutrients directly benefit human nutrition." Said Gregario, "If this new variety is successful, then micronutrient deficiency may be a part of history—like smallpox or polio. But that's still just a dream."

Golden Rice is also still just a dream—although Monsanto's Horsch confidently predicts that it will eventually reach the people it was designed to help. For Golden Rice inventor Ingo Potrykus, bioavailability studies are critical. He agrees that to get governments in developing countries and international humanitarian groups to back Golden Rice as a means of easing vitamin A deficiency, IRRI must prove that the beta carotene in the rice is usable and will make a difference. Until such tests are done, the developers of Golden Rice remain open to attacks from critics who believe they are making "false promises" by claiming that it will help alleviate vitamin A deficiency.

Golden Rice, said Greenpeace, is "fool's gold." Michael Pollan, in a *New York Times Magazine* article, cited a figure that appears to have come from Greenpeace's propaganda: "An 11-year-old would have to eat 15 pounds of cooked golden rice a day—quite a bowlful—to satisfy his minimum daily requirement of vitamin A." Yet this figure is merely a conjecture. Robert Russell, a nutritionist at Tufts University and a specialist in vitamin A nutrition, has calculated a figure of 200 grams per day—7 ounces—or two out of the three to four bowls full that an adult on a rice-based diet ordinarily eats. Which figure is right? It will take a bioavailability trial to learn. But as long as Golden Rice is confined to the greenhouse, the protocol established for the iron-rich IR68144 cannot be followed: enough rice cannot be grown under glass to feed 300 nuns for nine months.

Lecturing at Yale University in April 2003, Potrykus was asked if Golden Rice, like Borlaug's wheat and the other Green Revolution crops, would not simply contribute more to the problem of malnutrition by encouraging poor farmers to move from nutritious vegetable-based multi-cropping systems to a rice monoculture. "It was not without reason that production moved from high nutritious low-yield crops to low nutritious high-yield crops," answered Potrykus. "You can either die from hunger or from malnutrition. I don't know what is worse. The solution I am offering is to make high production crops like rice or wheat more nutritious."

Asked why he invented Golden Rice, he answered, "I have been asked this before, and I have thought about it. I'm a refugee myself, from a part of Germany generously given to Russia after the war." For Potrykus, a year shy of 70, "the war" is World War II. "I lost my father in the last days of the war," he continued. "He was a medical doctor. My mother had to raise four children without anything. She managed to allow us all higher education. But we have experienced hunger. Between the ages of 12 and 14, much of my brother's and my attention was given to where to find something to eat. You could say we reaped what we did not sow."

After earning a college degree in biology, Potrykus taught high school. He married and had children. He considers himself first "an old fashioned field biologist, a naturalist. From my mother I have this strong interest in nature," he said. A project he has been working on for many years is to videotape all the birds of North America. "I was teaching in high school when the director of the Max Planck Institute for Plant Breeding, Josef Straub, offered me the possibility of a Ph.D. I was exposed to a lot of breeding practices. I did the Ph.D. while still teaching half time, and then I went back to teaching full time. I came into this idea already when I was teaching high school. I was running courses then on the topic, 'More Food for More People.' This looks strange now, because it was the early sixties and the peak of the Green Revolution."

He was still teaching More Food for More People 20 years later at the university level. "In my student evaluations I heard again and again the complaint that I am talking about transgenic plants and they don't want to hear about it. I felt I had to produce a case that demonstrates to my students that they are wrong, that you can use the technology for a good purpose."

Does the world really need more food? Since 1798, when Malthus published his *Essay on the Principle of Population*, catastrophists have predicted imminent famine. Nearly 200 years later Paul Ehrlich, like Malthus, reduced the problem to its simplest form: too many people, not enough food. In his 1968 book *The Population Bomb*, Ehrlich concludes, "The battle to feed all humanity is over." "At this late date," he says, "nothing can prevent a substantial increase in the world death rate." Ehrlich was writing on the eve of the Green Revolution—the largest expansion in agricultural productivity in the history of human civilization. His prediction, like Malthus's, did not come true.

And yet the earth is finite.

In the course of just one century, the twentieth, the human population doubled twice: from 1.5 billion to 3 billion to 6 billion. It continues to expand by 80 million people a year. It doesn't take a catastrophist to see that humanity is pushing against some planetary limits. The demands of agriculture and industry, human habitation, and transportation, are driving to extinction more species per year than at any time since the Cretaceous. In 1998 Dan Simberloff, an ecologist at the University of Tennessee, was quoted in the *Washington Post*: "The speed at which species are being lost is much faster than any we've seen in the past—including those [extinctions] related to meteor collisions."

There's little doubt that the trends of the twentieth century cannot continue. In his 1995 book *How Many People Can the Earth Support?*, Joel Cohen lays out the problem of the earth's "carrying capacity" in all its complexity. There is no single answer, no simple answer to his title's question. The number of people Earth can accommodate depends on how they live and how well they manage the planet's physical, chemical, and biological environments. The choices available to us—and the choices we make—depend on science and technology, on the one hand, and on politics, preferences, and moral judgments, on the other.

There is no single, simple path to a sustainable future, either. To meet people's needs without further harm to the environment requires changes of many kinds. Cohen puts them into three categories: we must "put fewer forks on the table" (decrease population growth and reduce consumption), "make a bigger pie" (grow more food), and "teach better manners" (change how people interact with each other for everyone's benefit).

The rate of population growth *is* declining—more rapidly than the experts predicted even a decade ago, when the population was expected to double yet again before stabilizing. Some of the underlying trends are positive, such as improvements in education and economic development, particularly for women. But some are negative. They are the familiar scourges of too many people: disease, famine, war. As Garrett Hardin, author of the famous essay "The Tragedy of the Commons," points out, no one dies of overpopulation.

Experts now estimate that the number of people will stop growing by the middle of the twenty-first century. Before then, however, some 3 billion more people will be living on Earth than are alive now. This number is nearly 10 times the population of the United States today. Many—probably most—of these people will live in countries that are, even now, unable to provide their people with enough food for good health.

Putting fewer forks on the table doesn't just mean decreasing the population growth rate. It is also about how much we eat, what we eat, and what we waste. An adult needs between 2,000 and 2,200 calories per day—more for men and less for women, more for those who do heavy manual labor and less for those who sit at desks. In *Feeding the World*, Vaclav Smil estimates that the total food available per person in 1990, according to data from the FAO, was about 2,700 calories per day. Averaged worldwide, each person ate about 2,000 calories per day. The rest—about 700 calories per person per day—was lost during harvesting, processing, and distribution, or simply discarded.

These figures say that we already produce enough food to feed the world. What they conceal is the appalling gap between the richest and poorest nations. The amount of food each person in a rich country eats every day is at the high end of the range, about 2,200 calories each day—and sometimes even higher, up to 2,700 calories. In the poorest countries, people eat an average of 1,500 to 1,700 calories per day.

The amount of food available to people in rich and poor countries is also very unequal. According to FAO statistics, the per person food availability in the United States between 1992 and 1994 was about 3,600 calories per day, some 40 percent more than was eaten. (Available, in this case, assumes a person can afford to buy it, yet even in the United States millions of people use food stamps or go hungry.) In less developed countries, only 10 to 15 percent more food was available than was eaten. Moreover, the people in affluent countries consume between two and four times as much milk and meat as the world's average citizen. Converting plant foodstuffs, like grass and grain, to milk and meat is inefficient. Of the usable energy in animal feed, only 33 percent ends up as food energy in milk, 20 percent in pork, and 6 percent in beef.

But getting everyone to adopt a vegetarian lifestyle is not a likely solution, even if every grain and vegetable is biofortified. As Dennis Avery of the Hudson Institute points out, "No country or culture in history has voluntarily accepted a diet based solely on the relatively low-quality protein found in vegetable sources. Meat and milk consumption is rising by millions of tons per year in China and India right now as their incomes rise." Even among America's 12 million vegetarians, only 4 percent never eat any animal products, according to a survey commissioned by the *Vegetarian Times* in 1992. Fifty percent agreed with the statement, "In order to satisfy my appetite, a main meal must include meat." Nonetheless, Smil argues, cutting down the amount of food rich nations waste, reducing their meat consumption to a healthier 25 percent of total calories, and breeding animals that are more efficient in converting feed to food would go a long way toward feeding the world of the future.

Yet even if food is shared out fairly among all peoples—with no waste —we will soon need more: in 50 years, there might be another 3 billion people who need to eat. To "make a bigger pie," in Cohen's words, we have two choices. We can cultivate more land, knowing that land put under the plow is land taken away from black bears and monarch butterflies, Bengal tigers and tropical birds. Or we can produce more food from the land that is already being farmed.

The huge increases of the Green Revolution came from improvements in the yield of each plant, which made each farmed acre give more food. But in the last decade, yields of the major grains have not increased significantly. Agronomist Ken Cassman has argued that some crops are reaching their yield limits, the maximum amount of grain that they can produce under the very best of weather and fertilization conditions. Where will additional yield gains come from?

When Ingo Potrykus in Switzerland was lecturing his high-school students on the subject of More Food for More People, no one could begin to answer this question. What has changed since then is our knowledge of plant biology. Today we know the sequence of the entire genome of the tiny model plant *Arabidopsis thaliana*, as well as the genome sequences of two rice varieties. Work on sequencing the corn genome is underway, and a start has been made on the genomes of wheat, soybeans, and many other crops. Knowing the genome sequences makes it vastly easier to identify, analyze, change, and reintroduce genes that affect critical processes in plants. Comparing genomes, we've identified the similarities among genes and we've understood that what is learned in one plant can often be applied to another. All of these advances have quickened the pace of discovery and broadened our knowledge of photosynthesis, of how plants use nitrogen, and of how they cope with excess salt, toxic chemicals, and lack of water. Together with the ability to move genes between plants and into plants from other sources, this knowledge lets us begin tackling the barriers that limit agricultural productivity.

The first of these barriers is nitrogen use. Plants, together with the bacteria that live in and with them, convert carbon and nitrogen in the air into sugars and amino acids. These sugars and amino acids are the basic building blocks of the starches and proteins of which plants are made—and which feed both humans and their domestic animals. The amount of nitrogen that can be provided by bacteria or derived from composted animal and plant materials is much less than plants can use. Once people discovered how to fix nitrogen from the air, converting it into fertilizers, it was possible to overcome this limit on the nitrogen supply. Crop yields rose as fertilizer plants were built all over the world.

But beyond a certain point, adding more fertilizer no longer helps. Plants use only about half of the nitrogen applied as fertilizer, even under the best conditions. Much of the rest runs off with the rain into streams, rivers, lakes, and oceans. There it becomes a major pollutant. It acts as a fertilizer for small organisms, particularly algae, whose populations explode. Algae produce oxygen by day and consume it at night, depleting the oxygen available to other animals. When algae die, their decay also uses up oxygen in the water. The end result of an algal bloom is the suffocation of fish and other animals that live deeper down.

One way to solve the problem of nitrogen pollution is to increase the plants' ability to use nitrogen, turning more fertilizer into plant proteins and leaving less to run off the land. By increasing the crop's yield, such an improvement could also benefit farmers who do not use chemical fertilizers, either because they cannot easily afford them or because they are limited by the Organic Rule.

Whether it comes to the plant by means of nitrogen-fixing bacteria, from animal manure or plant compost, or from chemical fertilizer, plants first convert nitrogen to ammonia (NH_3). Then a plant enzyme called glutamine synthetase attaches the nitrogen atom in the ammonia to glutamic acid, an amino acid. Once attached to glutamic acid in the plant, the nitrogen can be moved by other enzymes into a variety of small molecules, including all of the other amino acids. These, in turn, are converted to proteins and other nitrogen-containing compounds. When researchers introduced a bean glutamine synthetase gene into wheat, providing more glutamine synthetase, the wheat plants developed faster, flowered earlier, and produced heavier seeds containing more protein.

Another important nitrogen enzyme is glutamate dehydrogenase. When nitrogen is abundant, glutamate dehydrogenase seems to help the plant redistribute it and make better use of its supply. The enzyme removes the nitrogen from amino acids, leaving keto-acids and NH_3 that are then available to make other needed compounds. Again, preliminary experiments show that plants containing an extra, highly expressed glutamate dehydrogenase gene grow bigger, yielding more biomass than plants that don't carry the extra gene. While these observations are still far from the wheat field, they suggest that changing the expression of a relatively small number of genes might produce substantial gains in crop yields.

Yields don't depend just on nitrogen. They also depend on how efficiently the crop plant makes use of carbon. Plants take carbon from the carbon dioxide (CO_2) in the air and convert it into a sugar molecule, which consists of carbon and the hydrogen and oxygen from water (H_2O). The energy for this reaction comes from photosynthesis. The plant absorbs light and uses it to increase the energy level of electrons. The excited electrons then trickle down through a chain of proteins. There, the energy is extracted and used to drive the reduction of CO_2 by the enzyme RuBP carboxylase, commonly called Rubisco—the most abundant protein on Earth. The net result of this reaction is that the carbon atom is incorporated into a sugar molecule.

Plants differ in their ability to capture CO_2 and in how efficiently they convert it into sugars. The basic photosynthetic process, called "fixing" carbon, captures the carbon in a sugar-like molecule that has three carbon atoms. Plants that can carry out only this basic reaction are called C_3 plants. But some plants, designated C_4 plants, have an additional pathway that makes a four-carbon sugar. In C_3 plants, photosynthesis is always coupled to photorespiration, which is carried out by the same enzyme, Rubisco, and drives the reaction in reverse. That is, photorespiration consumes oxygen and releases CO_2, but it doesn't capture the energy this reaction produces—that energy is wasted. As much as half of the carbon drawn from the air in the first place is released again through photorespiration, using up energy in the process.

Although C_4 plants use a bit more energy to fix carbon in the first place, overall they can be two or even three times as efficient as C_3 plants. C_4 plants have an additional enzyme, called PEP carboxylase, that fixes carbon into a four-carbon sugar. PEP carboxylase, unlike

Rubisco, is not bothered by oxygen. In simple terms, the C_4 photosynthetic pathway serves as a CO_2 pump. It concentrates CO_2 near the Rubisco enzyme, suppressing the oxygen-driven photorespiration in favor of carbon fixation. A majority of land plants, including rice, wheat, oats, and rye, are the less-efficient C_3 plants. Among major crops, only corn is a C_4 plant. Although many plant breeders have tried, they have not been able to transfer the C_4 traits to C_3 crop plants using conventional breeding techniques.

In 1999 Maurice Ku of Washington State University, together with a group of Japanese scientists at the National Institute of Agrobiological Resources in Tsukuba and at the BioScience Center of Nagoya University, reported that he had successfully transferred the PEP carboxylase gene from maize to rice. The researchers also tested a second maize gene, one that encodes pyruvate orthophosphate dikinase, an enzyme that provides one of the compounds that the PEP carboxylase uses in fixing carbon. The rice plants carrying either maize gene showed higher rates of photosynthesis. Oxygen didn't interfere with photosynthesis in the rice plants carrying the maize PEP carboxylase gene—this oxygen-insensitivity of carbon fixation is the hallmark of the C_4 plant. The gene was expressed at a high level. Indeed, as much as 12 percent of the protein in these plants was PEP carboxylase. The researchers reported that the yields of the rice plants that expressed the added corn PEP carboxylase were 10 to 20 percent higher that those of the parental plants. The rice plants expressing the other maize gene gave yields as much as 35 percent higher.

These results, while they await confirmation and are a long way from being applied to agriculture, make plant breeders optimistic. Molecular techniques might be able to break through limits that have long stymied their best efforts. Indeed, they make it possible to alter the fundamental biochemical reactions that set the upper limit on the yields of crop plants today.

Another limit on yield is water. Lack of water—drought—and too much salt (which has the same effect) reduce crop yields around the world. Both dehydrate plants: water from the inside of the plant's cells moves out by a process called osmosis. This loss of water triggers severe stress reactions. Photosynthesis shuts down. In extreme cases the plant dies. Salty soil can come from irrigating too heavily for too long. Much of the irrigation water simply evaporates, leaving behind whatever salts were dissolved in it. Soil salinization affects more than 700 million acres of otherwise arable land. If plants could better withstand salt stress, yields could increase even on marginal land.

Some plants have evolved to live comfortably under very salty conditions. These plants, called halophytes, have several mechanisms that prevent or limit the damage done by water loss and too much salt. Some plants pump out sodium, the most damaging component of salt. Others accumulate it inside of vacuoles, central compartments in their cells. Still others fill up their cells with compounds called osmoprotectants that keep the water inside. Sometimes these compounds are sugars or amino acids, but plants—and marine algae—use other compounds as well. A number of studies have shown that introducing genes that code for enzymes that produce osmoprotectants increases plants' ability to withstand salt stress.

So far, most of these studies have used genes that come from organisms other than plants, particularly bacteria. Plant genes have yet to be explored for this purpose. But both genes that encode salt pumps and those that code for enzymes that cause osmoprotectants to be made are being identified and studied. Other genes are being investigated as well, such as the regulatory genes that control the plant's overall response to salt.

Still another major factor that limits crop yields is the quality of the soil. Not all soils are hospitable to plants. One major problem in acid soils, for instance, is aluminum. Aluminum is the third most abundant element on Earth. When soil is alkaline or neutral, aluminum is in a form that doesn't harm growing plants. When the soil's acidity increases, the aluminum is converted to a soluble form that is toxic to plants. Soil acidification is exacerbated by some farming practices and by acid rain. It affects an estimated 40 percent of arable land worldwide. In the tropics, aluminum toxicity cuts yields by as much as 80 percent on about half of the arable land. Even at quite low concentrations, aluminum ions inhibit root growth, which in turn affects plant growth and yield.

Plants have developed several mechanisms for aluminum tolerance. Some plants exclude it. Others secrete organic acids, such as citric acid, oxalic acid, and malic acid, that bind tightly to the aluminum and prevent the plant from absorbing it. It is the growing root tips that secrete the acids, forming a protective shield. In 1997 a research group led by Luis Herrera-Estrella at the National Polytechnic Institute in Irapuato, Mexico, reported that the ability of a papaya plant to tolerate aluminum could be enhanced by introducing a bacterial gene coding for citrate synthase, the enzyme that produces citric acid. The plants produced and secreted more citric acid, allowing them to grow in soils that had been toxic to them previously. These experiments establish the principle of using genes to enhance the aluminum tolerance of plants, but the first experiments might not provide the final answers. As more is learned about how plants tolerate aluminum, more genes will be identified.

Limits on nitrogen and carbon use, salt and aluminum toxicity— these are among the major problems that must be overcome if farmers' yields are to double or perhaps even triple to meet the demands of a human population 8 or 9 billion in number and demanding more and better food. It seems unlikely that the future holds another simple breakthrough, like the synergy between dwarfing genes and fertilizer that made the Green Revolution possible. But a breakthrough that enhances either the use of nitrogen or the efficiency of photosynthesis, or both simultaneously, could push yields up dramatically. More likely, the advances will be incremental. Small improvements of many different kinds will be made in many different crops. But it depends. And what it depends on has rather less to do with the science than with people.

❧

In 2002 Zambia's president rejected a shipment of donated corn from the United States, ostensibly because genetically modified food had not been proven safe to eat. According to the *Los Angeles Times*, "Many Zambians in rural areas have resorted to eating leaves, twigs, and even poisonous berries and nuts to cope with the worst food crisis in a decade hitting southern Africa." Zambian president Levy Mwanawasa had declared a food emergency in the nation three months earlier. Yet he refused the American maize, saying, "We would rather starve than get something toxic."

His choice was bewildering. The health consequences of starvation are unequivocally terminal—and there is no evidence that genetically modified corn is toxic. As the *Los Angeles Times* reported, "The United States, United Nations, and humanitarian aid groups insist that the U.S.-donated corn is safe and identical to grain eaten daily by people in the United States, Canada, and other countries."

Mwanawasa's logic is indeed incomprehensible—unless one views it from an economic standpoint. The European Union has urged African governments who want to trade in Europe to treat genetically modified crops as a serious biological threat. If the Zambian government were to lose its "GM-free" status, it would lose access to European markets, where its exports include organic baby corn and carrots. And indeed, President Mwanawasa was quoted as saying that he does not want the introduction of genetically modified foods to hurt his export trade with Europe. The government of Zimbabwe, also facing famine in 2002, agreed to accept U.S.-donated corn only if it was first ground into cornmeal "so that the food aid cannot be planted," the BBC reported. "Zimbabwe and some of its neighbors are worried that GM seeds could contaminate locally grown crops, threatening lucrative exports to Europe, which insists that food must be GM-free."

"All across Britain and most of the rest of Europe," the *New York Times* reported in February 2003, "shoppers would be hard pressed to find any genetically modified, or GM, products on grocery store shelves, and that is precisely how most people want it." At the Happy Apple greengrocer in the small English town of Totnes, "the roasted vegetable pasty is labeled, clearly and proudly, as GM-free," the *Times* reported. When asked her opinion on GM foods, one shopper replied, "It's a kind of corruption, not the right thing to do, you know?"

The private and personal choices of European shoppers like this one are setting the public policy of African nations. Zambia's decision to refuse American corn was greeted with disbelief around the world. But a thoughtful look reveals it to be a logical, if unintended, consequence of the expression of a preference on the part of Europeans for foods that have not been modified by molecular techniques. African and other less developed nations are caught in a terrible bind. With almost three million people at risk of starvation, they are faced with a choice between immediate suffering and closing the door on future economic prosperity.

The heart of economic development is the ability to grow more crops than growers and their families need to survive. The Rockefeller Foundation's Gordon Conway and Gary Toenniessen note that two-thirds of sub-Saharan Africa's more than 600 million people live on small farms. The food they produce, combined with what they can afford to buy, is insufficient. The result is that 194 million Africans, mostly children, are undernourished. "Africa does not produce enough food to feed itself even with equitable distribution," they wrote in 2003. "Food aid to Africa—currently running at 3.23 million tons annually—helps prevent starvation but can create dependency."

The first step, they say, is to achieve food security, which simply means reliable access to enough food to lead a healthy, active life. "Most African farmers have land assets adequate to provide food security and to rise above subsistence." But to do so, "they need to intensify production with genetic and agro-ecological technologies that require only small amounts of additional labor and capital." In a July 2003 *New York Times* Op-Ed piece on the same subject, Norman Borlaug argues: "Biotechnology absolutely should be part of African agricultural reform; African leaders would be making a grievous error if they turn their backs on it." He strongly urged African leaders not to follow the lead of Europe, where biotechnology has been "demonized," but to use it for the benefit of their farmers and their people.

How can Africa consider adopting molecular technology if by doing so its farmers are locked out of European markets? How can Africa afford not to adopt approaches that are biology-based, low-cost, and beneficial on both small and large scales? How did we—the industrialized nations that developed these molecular techniques for plant breeding—contribute to this extraordinary and deeply distressing state of affairs?

Calestous Juma, director of the Science, Technology and Globalization Project at Harvard University's Kennedy School of Government, compares it to the persecution of coffee. "In the 1500s," he explains, "Catholic bishops tried to have coffee banned from the Christian world for competing with wine and representing new cultural as well as religious values." Juma continues, "In public smear campaigns similar to those currently directed at biotech products, coffee was rumored to cause impotence and other ills and was either outlawed or its use restricted by leaders in Mecca, Cairo, Istanbul, England, Germany, and Sweden. In a spirited 1674 effort to defend the consumption of wine, French doctors argued that when one drinks coffee: 'The body becomes a mere shadow of its former self; it goes into a decline, and dwindles away. The heart and guts are so weakened that the drinker suffers delusions, and the body receives such a shock that it is as though it were bewitched.'"

In coffeehouses throughout Europe and, increasingly in America, similar campaigns are being waged now against genetically modified foods—using equally exaggerated claims of potential harm. Juma writes: "Debates over biotechnology are part of a long history of social discourse over new products. Claims about the promise of new technology are at times greeted with skepticism, vilification, or outright opposition—often dominated by slander, innuendo, and misinformation. Even some of the most ubiquitous products endured centuries of persecution."

It is sobering, for example, to recollect that vaccinations against smallpox—a disease that kills 30 percent of the people it infects and disfigures the rest—were vilified in editorials and cartoons, publicly protested, and strongly resisted. Fortunately, national governments and the United Nations persisted in vaccinating people—sometimes even with a bit of coercion—and smallpox is gone.

The problem today, suggests Stanford University's Henry Miller, a former FDA official, is compounded by governments that increasingly depend on public opinion in formulating policy involving scientific issues. Noting that in 2003 the British government organized focus groups "to find out what ordinary people really think [about GM foods] once they've heard all the arguments," Miller says: "Getting policy recommendations on an obscure and complex technical question from groups of citizen nonexperts (who are recruited through newspaper ads) is similar to going from your cardiologist's office to a café, explaining to the waitress the therapeutic options for your chest pain, and asking her whether you should have the angioplasty or just take medication."

Even when they've "heard all the arguments," intelligent and inquiring minds not trained in the subject can still be confused. In June 2003, for instance, a Zambian newspaper reported that "Maize is not directly consumed in America." The writer, Simon Mwanza, was one of seven African journalists who had toured the United States to learn about biotechology, for which he used the abbreviation BT. His tour had included stops at Monsanto and Pioneer Hi-Bred, several universities, the Center for Science in the Public Interest, the Pew Initiative on Food and Biotechnology, the National Corn Growers' Association, USAID, and the Washington office of a senator from Iowa. None, apparently, had invited him to try cornflakes, corn chips, or corn-on-the-cob. Or perhaps the problem was one of translation, and the fact that "corn" was the American name for "maize" was not made plain. At the University of Maryland he learned that "most of the maize produced in the USA was for animal feed," he wrote, and that "the US also uses maize to produce ethanol." Later, visiting the National Corn Growers Association, he wrote, "The journalists' eyes popped out when they were shown a wide range of products made from maize—thanks to BT." The Association representatives gave the visiting journalists "t-shirts made from corn to make their BT point abundantly clear." Mwanza's conclusion about biotechnology? "While BT appears a promising solution to agriculture, it is difficult to forget what Dr. Scott Angle of Maryland University said: 'We don't know what we don't know.'"

Most people have not devoted even two weeks, as Mwanza and his colleagues did, to understanding the technology behind genetically modified foods. Still, they have strong opinions. A 2003 survey in America found that 58 percent of the people asked were "unwilling to eat genetically modified (GM) food." That majority response seems to send a clear signal to food producers and seed companies. That is, until you ask the next question: "What food are Americans willing to eat?" A 1993 survey of New Jersey residents found that 41 percent of the respondents would not eat food produced through "traditional hybridization techniques." A full 20 percent said it was "morally wrong to produce plants this way." Yet since 1970, more than 96 percent of the corn grown in America has been hybrid corn. Sweet corn, corn flakes, muffins, and chips, the corn fed to our beef cows, chickens, and pigs is all hybrid corn, as is the source of the corn sweeteners and corn starch found in mayonnaise, peanut butter, chewing gum, soft drinks, beer, wine, frozen fish, processed meats, all dehydrated foods, all powdered foods, and all granulated foods.

It is perhaps not surprising that the Organic Rule, so heavily influenced by public opinion, forbids the use of irradiation, antibiotics, and molecular genetic modifications in producing "organic" food. But in the end, says Miller, "The goal of policy formulation should be to get the right answers." "Although it may be useful, as well as politic, for governments to consult broadly on high-profile public policy issues," he adds, "after the consultations and deliberations have been completed, government leaders are supposed to *lead*."

Getting the right answers on genetically modified foods matters—profoundly. The science is complex, and advancing daily. As we continue to learn more about how plants grow, and as we become more skillful in transferring useful genes into plant cells, we find ourselves with an opportunity to get it right—not just for the economic benefit of large companies, but for the benefit of ordinary people everywhere in the world.

Getting it "right" will have many local meanings. It will mean virus-resistant tomatoes in Italy, herbicide-resistant wheat in Washington, and insect-resistant Bt corn in Iowa. It will mean aluminum-tolerant crops in the tropics and virus-resistant sweet potatoes in Africa. Some of these crops will be produced by companies because they can return a profit, every company's prerequisite for survival. But others might never make money. These will come only when governments everywhere recognize—and invest much more heavily in—agriculture both as a public good and an environmental necessity. As well, these will come only when regulators and regulations become more responsive to evolving knowledge than to public perceptions and anxieties. Only then will public sector scientists be able to invest their time and knowledge in raising yields in an ecologically sound way.

And yet, in the deepest sense, getting it "right" is the same for all nations: having enough to eat while preserving and protecting the environment. Every civilization rests on food. Over many thousands of years, humans have devoted a great deal of intellect, energy, and effort to changing wild plants into food plants. These changes—all of them—involved changes in the plants' genes. We have a long history of tinkering with nature. It is no exaggeration to say that our tinkering, our modification of plant genes—and those of domesticated animals—to meet our nutritional needs, has shaped our world.

At the same time agriculture in its very essence is ecologically destructive, whether it is performed at the subsistence level for a single family or on an industrial scale. The challenge now, as our population pushes against the planet's limits, is to lessen the destructive effects of agriculture on the earth even as we coax it to produce more food.

"To assert that GM techniques are a threat to biodiversity is to state the exact opposite of the truth," writes Peter Raven, director of the Missouri Botanical Garden. "They and other methods and techniques must be used, and used aggressively, to help build sustainable and productive, low-input agricultural systems in many different agricultural zones around the world."

At the International Botanical Congress in 1999, Raven announced: "We are predicting the extinction of about two-thirds of all bird, mammal, butterfly, and plant species by the end of the next century, based on current trends." In a 2003 essay he elaborates, "These organisms are simply beautiful, enriching our lives in many ways and inspiring us every day. By any moral or ethical standard, we simply do not have the right to destroy them, and yet we are doing it savagely, relentlessly, and at a rapidly increasing rate, every day. Many believe, and I agree with them, that we simply do not have the right to destroy what is such a high proportion of the species on Earth. They are, as far as we know, our only living companions in the universe."

And the greatest danger to other species is our own need for food. "Nothing has driven more species to extinction or caused more instability in the world's ecological systems than the development of an agriculture sufficient to feed 6.3 billion people," Raven says. "The less focused and productive this agriculture is, the more destructive its effects will be."

Using our growing knowledge of plants and plant genes, and our increasing skill at modifying them with molecular techniques, we can make agriculture more focused and more productive—if we are careful. The thoughtful choice of genetic modifications can help us become better stewards of the earth. The key words here are careful and thoughtful. Whether the technology will be helpful or harmful, in the long run, depends on how it is used, on the choices people make. The better we understand what this technology is—how it has come to be and what it involves—the wiser will be the decisions we make as a civilization about how it will be used in the future.

ACKNOWLEDGMENTS

The authors would like to thank the following people for contributing to this book by agreeing to be interviewed, by responding to queries, by providing scientific papers and background material, and/or by reading and commenting on parts of the manuscript:

Klaus Ammann, Chuck Armstrong, Jim Astwood, Gerard Barry, Roger Beachy, John Beard, Ken Cassman, Herb Cole, Jim Cook, Richard Craig, John Doebley, Eric M. Hallerman, Rob Horsch, Ernie Jaworski, Calestous Juma, Lee Kass, Lawrence Kent, Avraham Levy, Henry Miller, David Mortensen, Per Pinstrup-Andersen, Steve Padgette, Ingo Potrykus, John Purcell, Peter Raven, Eric Sachs, Karel Schubert, Pat Shipman, Nigel Taylor, Gary Toenniessen, and Catherine Woteki.

NOTES

1: AGAINST THE WAYS OF NATURE

Ingo Potrykus was interviewed by Nancy Marie Brown in April 2003; publications by Potrykus and his colleagues (particularly those listed in the bibliography as Beyer et al. (2002), Lucca et al. (2001), and Ye et al. (2000)) were major sources for this chapter. Mary Lou Guerinot of Dartmouth College wrote the "glowing commentary." Additional sources include publications by and interviews with Gary Toenniessen and Klaus Ammann (who were interviewed by Brown in February 2003) and correspondence with Henry Miller. Much of the backlash against Golden Rice, and many of Potrykus's responses to his critics, are posted on www.agbioworld.org. C. K. Rao's "rice, green chilies, and salt" comment was printed by Jan Bowman in *Spiked*, 19 February 2003 (www.spiked-online.com). Datta's comments and information on the current status of Golden Rice can be found at www.irri.org.

Good sources for the history of plant breeding are Kloppenburg (1988), Lurquin (2001), and Torrey (1985). Background information on rice came from Te-Tzu Chang's chapter in Kiple (2002); Khush (1977); and Oka (1988). Richard Langer speculated on Versace's taste in cauliflower in the *New York Times*, 20 February 2002. The All-Red Potato is described by Weaver (2000). Information on Rio Red grapefruit came from Texas A&M University's Citrus Center and on Above wheat from the Colorado Agricultural Experiment Station. All quotations concerning Burbank and his work are from Dreyer (1985).

The history of mutation breeding, including the colchicine "craze," can be found in Harten (1998). The description of triticale as the first "intergeneric hybrid" is from Simmonds (1976). Raven's opinion of protoplast culture comes from the fourth edition of *Biology of Plants* (Worth Publishers, 1986). T. Kinoshita and K. Mori's protoplast fusion techniques are described in Maluszynski and Kasha (2002), as are W. Navarro Alvarez's salt-tolerant rice experiments, B. P. Forster's comments on Golden Promise barley, and the anther culture experiments of Q. F. Chen, C. L. Wang, Y. M. Lu, M. Shen, R. Afza, M. V. Duren, and H. Brunner. Creso wheat is discussed by P. Donini and A. Sonnino in Jain et al. (1998); in the same volume, P. C. Remotti comments on the nonuniformity of plantlets grown from callus tissue. The Mutant Variety Database was established by the International Atomic Energy Agency and the Food and Agriculture Organization of the United Nations (FAO) in 1969 and is available on the Internet at www-infocris.iaea.org/MVD/nav.htm.

The coining of "Frankenfood" is discussed by William Safire in the 13 August 2000 *New York Times*; thanks to Eric Hallerman for the reference.

2: THE WILD AND THE SOWN

The beginnings of agriculture are discussed by E. Anderson (1967); P. C. Anderson (1999); J. Cohen (1995); M. N. Cohen (1977) and in his chapter in Kiple (2002); Diamond (1999 and 2002); Farb and Armelagos (1980); Fernandez-Armesto (2002); Harlan (1995); Heiser (1990); Jones (1969 and 1995); Kloppenberg (1988); MacNeish (1991); Oka (1988); Salvador (1997); Simpson and Ogorzaly (1995); Smith (1995 and 2001); Tudge (1995 and 1998); and Zohary and Hopf (1993). Florence Shipek summarizes her investigations of the Kumeyaay in Harris and Hillman (1989). Early attitudes toward potatoes, corn, chocolate, coffee, tomatoes, and other crops are noted by Coe and Coe (1996), Evans (1998), Fussell (1992), Pendergast (1999), Visser (1986), and Zuckerman (1998); by Timothy Johns and J. G. Hawkes in Harris and Hillman (1989); and by Jonathan Beecher Field in Benes (1996). The early settlers' dislike of tomatoes is chronicled by Andrew F. Smith in Benes (1996), while Daniel A. Romani Jr. in the same volume discusses the importing of grasses and clover.

Recent studies of wheat include those by Buckler et al. (2001); Huang et al. (2002); Kellogg (2001); Salamini et al. (2002); and Daniel Zohary in Anderson (1999). Jack R. Harlan's stories of harvesting wild grass seed can be found in Anderson (1999) and Harris and Hillman (1989). Gordon C. Hillman and M. Stuart Davis reported on their wild wheat experiments in Anderson (1999).

Doebley and his colleagues have published their work on the origins of corn in a series of papers, including those listed in the bibliography as Bennetzen et al. (2001), Doebley (1992, 1994, and 2001), Doebley et al. (1997), Eyre-Walker et al. (1998), Jaenicke-Despres et al. (2003), Lauter and Doebley (2002), Matsuoka et al. (2002), Wang et al. (1999), and White and Doebley (1998). East's "A Chronicle of Corn" appeared in *Popular Science Monthly* in 1913. Mangelsdorf and Reeves published their theory in 1938 in the *Proceedings of the National Academy of Sciences*. Other details concerning the history of corn come from George W. Beadle's "The Ancestry of Corn," in *Scientific American*, January 1980, as well as from Iltis (1983 and his 2002 conference paper abstracted at www.2002.botanyconference.org/section4.abstracts/4.shtml); the description of Iltis is from Fussell (1992). The worldwide success of corn is noted by Michael Pollan in "When a Crop Becomes King," *New York Times*, 19 July 2002. Benz (2001) and Piperno and Flannery (2001) discuss the redating of the earliest known corncobs.

3: THE POWER IN THE EARTH

The theories of Malthus and Darwin, and the experiments of Priestley and Mendel can be found in many textbooks; it was Bernard D. Davis (1991) who suggested that the early lack of interest in Mendel's theories were due to his "mathematical approach." Sachs's *Lectures on the Physiology of Plants* (The Clarendon Press, 1887) is more obscure; Sachs's artificial plants were referenced by Torrey (1985). The early experiments on the nitrogen question, the influence of Liebig, and the comments of Wöhler on the vitalist principle are described in Evan Pugh's letters and other documents preserved in the archives of the

Pennsylvania State University libraries. Murray's career is described by Peter Childs in *Chemistry in Action*, Autumn 2000. Bateson's comments are cited by Carlson (1966).

Burbank and Shull's personality clash is described by Dreyer (1985). Pendergast (1999) comments on Sanka. Hubbell (2001) and Pollan (2001) tell the story of Johnny Appleseed. Additional information on grafting comes from Rayle and Wedburg (1975), while Kloppenberg (1988) discusses plant-gathering expeditions.

Shull's 1909 paper on the "pure-line method" was published in the *Report of the American Breeders' Association.* The response to hybrid corn is described by Kloppenberg (1988), Heiser (1990), and Duvick (personal communication, 2003), while the standards enforced by the corn shows are outlined in Henry A. Wallace and Earl N. Bressman's *Corn and Corn Growing* (John Wiley and Sons, 1949).

Norman Borlaug's biography derives from Conway (1997), from Gregg Easterbrook's profile in the *Atlantic Monthly* (January 1997), and from Rayle and Wedburg (1975), as well as from the Nobel Institute's e-Museum (www. nobel.se), on which two of Borlaug's lectures are also available. Borlaug's comments on Ghurdev Khush's achievements were published on www.irri.org.

The genetic mutation behind miracle rice is described in Sasaki et al. (2002).

4: GENES AND SPECIES

Ernst Mayr's classic book, *Systematics and the Origin of Species*, was published in 1942, yet how to define the term "species" and how to categorize plants remain active areas of research for taxonomists. Morrison (1998) discusses the problem of classifying wheat and its relatives; Morrison was also a coordinator of the "Wheat Synonymy" project at the USDA. In Hey (1997), a geneticist at Rutgers University tackles the species question. Interesting details concerning rice and the brassicas come from Gary W. Crawford's chapter in Cowan and Watson (1992); Heiser (1990); Oka (1988); Roberts (2001); Simmonds (1976); Simpson and Ogorzaly (1995); Tudge (1995 and 2000); and Weaver (2000).

Bud (1993), Carlson (1966), Davis (1991), and Portugal and Cohen (1977) discuss the early history of the concept of the gene. Biographical details about T. H. Morgan come from Allen (1975); Carlson (1966); Shine and Wrobel (1976); and Alfred Sturtevant's entry on Morgan in *Biographical Memoirs* 33 (1959). Judson (1979) cites Max Delbrück's comment that DNA was "stupid."

Bhattacharyya et al. (1990) identified the genetic basis of Mendel's wrinkled-seed trait.

5: TINKERING WITH EVOLUTION

The first complete sequencing of the human genome was announced in February 2001 simultaneously in *Science* and in *Nature*; the scientific papers were accompanied by numerous commentaries, including the essays mentioned here by Svante Pääbo in *Science* and David Baltimore in *Nature*. Jacob (1977) articulates the concept of evolutionary tinkering, which is further developed by Duboule and Wilkins (1998).

Studies and reviews on antifreeze genes in fish and plants include Cheng (1998), Huang et al. (2002), and Yeh et al. (2000). Dunsmuir's tomato experiments are described in Hightower et al. (1991). Martineau (2001) tells the story of Calgene's FlavrSavr tomato. Jennie Addario reported on the rally in which a protestor wore a tomato-fish costume in the Spring 2002 issue of the *Ryerson Review of Journalism*.

The McClintock biography mentioned is Evelyn Fox Keller's 1983 book, *A Feeling for the Organism*; the scientist whose experience contradicts Keller's description is Nina Fedoroff, co-author of the present volume, who worked with McClintock for several years and has written extensively on McClintock's discovery, as well as on her own work on transposons. Lee Kass of Cornell University was consulted for some biographical details concerning McClintock.

That Watson and Crick's description of the double helix was "tight as a sonnet" is the observation of Tudge (1995).

6: MAKING A CHIMERA

The history of canola oil comes from the chapter by Paul R. Mayers, Peter Kearns, Karen E. McIntyre, and Jennifer A. Eastwood in Atherton (2002); from Bill Wagner's article in the *FDA Consumer* magazine, November 1993; and from several reports on the website of Agriculture and Agri-Food Canada. Kneen's book was reviewed by *The Ottawa Citizen,* 5 July 1999.

The first DNA chimera is reported by Cohen et al. (1973). Cohen tells the story less formally in the July 1975 *Scientific American;* another version, his 1992 lecture "From Corned Beef to Cloning," is posted on www.accessexcellence.org.

Bud (1993), Conway (1997), Judson (1979), Lurquin (2002), and Russo (2003) all describe a "birth of biotechnology." Additional information about Lederberg and bacterial sex can be found on the websites of the National Library of Medicine and the Nobel Institute. The Nobel e-Museum (www. nobel.se) also provides information about Berg's recombinant DNA experiments, which were published as Jackson et al. (1972). Hardy's comments about transgenic cheese can be found on the Senate hearing website: http://www.senate.gov/~agriculture/Hearings/Hearings_1999/har99106.htm.

The spinach DNA experimental protocol by Teri Curtis is available on www.accessexcellence.org.

PCR is described by Mullis (1998) and Rabinow (1996). Shawn Carlson published "PCR at Home" in the July 2000 issue of *Scientific American.*

Torrey (1985), Busch et al. (1991), Lurquin (2001 and 2002), and Jack Widholm's chapter in Nelson (2001) summarize the techniques used to transform plants, as do Charles S. Gasser and Robert T. Fraley in the June 1992 *Scientific American.* Chilton et al. (1977), Schell and Van Montagu (1977), and Zambryski et al. (1980 and 1983) report the *Agrobacterium* method; Chilton describes it less formally in *Scientific American,* June 1983. Sanford (1988 and 2000) describes the "cowboy approach," or biolistics. The genome of *Agrobacterium* was described by Lurquin (2000) and *Genome News Network* (http://gnn.tigr.org/), 21 December 2001; the sequence is reported in Wood et al. (2001).

7: THE PRODUCT OR THE PROCESS

Major sources for this chapter include interviews with Roger Beachy, Lawrence Kent, Ernie Jaworski, Karel Schubert, and Nigel Taylor of the Donald Danforth Plant Science Center, and with Chuck Armstrong, Jim Astwood, Rob Horsch, Steve Padgette, John Purcell, and Eric Sachs of the Monsanto Company; the interviews were conducted by Brown and Fedoroff in November 2003.

The story of the virus-resistant papaya comes from the September 1998 APSNet Feature by Dennis Gonsalves, Steve Ferreira, Richard Manshardt, Maureen Fitch, and Jerry Slightom on the American Phytopathological Society's website (www.apsnet.org), as well as from Gianessi et al. (2002), Health Canada (www.agbios.com), and Cornell University (www.nysaes.cornell.edu). Post-transcriptional gene silencing in papayas was explained by Tennant et al. (2001).

The question of plant patents is covered by Ambrosoli (1997); Dreyer (1985); and Magnus, Caplan, and McGee (2002), particularly the chapters by A. M. Chakrabarty, R. S. Eisenberg, and J. Wilson. Background information was supplied by Richard Craig of the Pennsylvania State University, who was interviewed by Brown in June 2003. Maan's 1977 patent for hybrid wheat is available at http://patft.uspto.gov.

Regulation of DNA experiments and of genetically modified food is discussed by Berg et al. (1974), Kelman et al. (1987), Davis (1991), Turney (1998), Smith (2000), Martineau (2001), Paarlberg (2001), Atherton (2002), and Lurquin (2002). The relevant publications by the FDA are available at www.fda.gov. Fedoroff was a member of the NIH Recombinant DNA Advisory Committee from 1980-1984 and a member of the NIH Recombinant DNA Advisory Committee Working Group on Guideline Revisions in 1987. Fredrickson's and Stone's recollections are online at http://www.nih.gov/news/ and http://profiles.nlm.nih.gov/. Miller's comments were conveyed by personal correspondence; his letters and op-ed pieces have been published in the *Scientist* (8 July 2002, 30 September 2002, and 28 October 2002), on TechCentralStation.com (22 October 2002), the *European Wall Street Journal* (1 July 2003), and *Nature Biotechnology* (December 2003). Current regulations are detailed on the websites of the USDA,

FDA, and EPA. Statistics concerning the acreage planted to GM varieties comes from the International Service for the Acquisition of Agri-
biotech Applications, www.isaaa.org.

8: IS IT SAFE TO EAT?

Good sources of information on the safety of genetically engineered foods are the 2003 position paper of the Society of Toxicology
(http://toxsci.oupjournals.org/) and the July 2003 report of the U.K.'s
GM Science Review Panel, formed by Margaret Beckett (http://
www.gmsciencedebate.org.uk/). Michael Pariza addresses the same issues in a chapter in Kiple (2002), as do Jan Pedersen, Folmer D. Eriksen,
and Ib Knudsen in Custers (2001).

Per Pinstrup-Andersen was interviewed by Brown in February
2003; the international survey mentioned is discussed in Pinstrup-
Andersen and Schioler (2000). Petra Frey of the University of California wrote the teaching module; her protocol for extracting DNA from
tomatoes is available at http://ucbiotech.org/edu/edu_aids/
TomatoDNA.html. The calculation of up to a gram of DNA per day
comes from *Fremde DNA im Säugersystem* by W. Doerfler and R.
Schubbert (1997), as cited in the 2000 FAO-WHO report, "Safety Aspects of Genetically Modified Foods of Plant Origin."

Studies of the digestion of DNA include Doerfler (2000), Duggan
et al. (2000), and Schubbert et al. (1998). Gilbert's most recent work is
Netherwood et al. (2004); his burger and shake study is posted at
www.food.gov.uk/multimedia/pdfs/gmnewcastlereport.PDF.

Roelofs and Van Hasstert (2001) discusses the similarity of human
genes and bacterial genes. Brown (2003) examines horizontal gene
transfer.

Chambers et al. (2002) addresses the fate of antibiotic resistance
marker genes. The low probability of resistance passing between bacteria is discussed in the 2000 and 2001 FAO-WHO reports, the 2003
Society of Toxicology paper, the 2000 report to Congress cited as Smith
(2000), the American Medical Association's 2000 report (see www.ama-
assn.org), and the 2003 report of the U.K. GM Science Review Panel
(see www.gmsciencedebate.org.uk/). Sandeep Kumar and Matthias
Fladung describe new methods that do not use antibiotic marker genes

in Jain et al. (2002), as do Henry Daniell, in the same volume; Jack Widholm in Nelson (2001); and Lurquin (2002).

Kohli et al. (1999) was the CaMV 35S promoter study that sparked the response by Ho et al. (1999); Michael Hansen of Consumers Union elaborates in Nelson (2001). One of the authors of Kohli et al. (1999) is P. Christou, who rebutted Quist and Chapela's study of criollo corn (see Chapter 11). Basic information on cauliflower mosaic virus comes from *Central Coast Agriculture Highlights*, published by the University of California at Davis, and Shepherd's *Descriptions of Plant Viruses*, online at www3res.bbsrc.ac.uk/.

Information on the naturally toxic chemicals in foods comes from Ames and Gold (1997 and 2000); Farb and Armelagos (1980); Harlan (1995); Paarlberg (2001); Michael Pariza in Kiple (2002); and Visser (1986). The history of canola oil is from the chapter by Paul R. Mayers, Peter Kearns, Karen E. McIntyre, and Jennifer A. Eastwood in Atherton (2002); the same authors define "substantial equivalence." Roger Beachy and several colleagues discuss substantial equivalence in practice in the letter "Divergent Perspectives on GM Food" in *Nature Biotechnology*, December 2002.

Information on conventional plant breeding trials can be found at www.all-americaselections.org. Herb Cole of the Pennsylvania State University was interviewed about the Lenape potato by Brown in December 2003. Colorado State University's *Transgenic Crops: An Introduction and Resource Guide*, by Pat Byrne, Sarah Ward, Judy Harrington, and Lacy Fuller, is online at www.colostate.edu/programs/lifesciences/TransgenicCrops/history.html.

Levy's work with distant crosses was described in Kashkush et al. (2002). A collection of T-DNA insertions in *Arabidopsis* was described by Ichikawa et al. (2003).

Fedoroff was a member of the 2003 committee to review the FDA's guidelines concerning transgenic crops; she was also an external advisor to the FDA committee that initially approved the FlavrSavr tomato.

9: POISONED RATS OR POISONED WELLS

The Royal Society's review of Pusztai's work, "Review of Data on Possible Toxicity of GM Potatoes," is published in *Promoting Excellence in Science* (www.royalsoc.ac.uk); Pusztai's comments are posted at http://www.wsws.org/articles/1999/feb1999/food-f17.shtml; his *Lancet* paper is Ewen and Pusztai (1999); and his data are available at www.rowett.ac.uk/gmo/ajp.htm. M. Enserink followed the situation in a series of editorials for *Science* in 1998 and 1999; see volumes 281 (page 1184), 283 (page 1094), and 286 (page 656). The idea that the difference Pusztai reported was due to somoclonal variation is supported by Bhat and Srinivasan (2002). Kozekue et al. (1999) discusses the inheritance of glycoalkaloids in potatoes, and Patel et al. (2002) discusses their effects.

Research and reviews related to lectins can be found in Batelli (1997), Carlini and Grossi-de-Sa (2002), Down et al. (2000 and 2001), Fitches et al. (2001 and 2002), Foissac et al. (2000), Hester and Wright (1996), Naughton et al. (2000), Olsnes and Kozlov (2001), Pusztai et al. (1990 and 1993), Rao et al. (1998), Romeis et al. (2003), and Rudiger et al. (2000). The kidney bean lunch is reported in Freed (1999).

Allergenicity is discussed by Buchanan (2001); Kleter and Peijnenburg (2002); Dean D. Metcalfe in Atherton (2002); Marion Nestle in Kiple (2002); Earle Nestmann, Todd Copeland, and Jason Hlywka in Atherton (2002); and Taylor (2003). Heiser (1990) notes that Kellogg called peanuts "the noble nut." Gina Kolata wrote about the allergy-limiting drug in the *New York Times*, 11 March 2003. The "Fruit Detective" was profiled by John Seabrook in the 19 & 26 August 2002 issue of the *New Yorker*. Catherine Woteki of Iowa State University was interviewed by Brown in February 2003; John Beard of the Pennsylvania State University was interviewed by Brown in July 2003. Lehrer's comment was cited by Pinstrup-Andersen and Schioler (2000). Background information on current allergenicity testing was provided by Jim Astwood, who was interviewed by Fedoroff and Brown in November 2003.

Nestle's editorial (1996) accompanied publication of Nordlee et al. (1996), the results of the safety tests Taylor and his colleagues performed on the Brazil-nut protein-containing soybean. Cook is quoted in Smith (2000).

The story of StarLink corn is summarized in the 2003 report of the Society of Toxicologists. Betz et al. (2000); Brian A. Federici in Atherton

(2002); and Siegel (2001) discuss the safety of Bt. The CDC report is available at www.cdc.gov/nceh/ehhe/Cry9cReport/default.htm. Links to relevant news stories can be found on the Colorado State University website mentioned in Chapter 8, www.colostate.edu/programs/lifesciences/TransgenicCrops/history.html.

Federici's chapter in Atherton (2002) discusses the reduction in fungal toxins in Bt corn. Bob Buchanan described his work in testimony before the U.S. Senate Committee on Agriculture, Nutrition, and Forestry in 1999. Samuel Lehrer presented his work at the American Association for the Advancement of Science's annual meeting in Denver in 2003, which Brown attended. Eliot Herman's work is discussed in the 2003 Annual Report of the Donald Danforth Plant Science Center in St. Louis.

10: THE BUTTERFLY AND THE CORN BORER

"The Butterfly Flap," as journalist Peter Pringle named it in *Prospect Magazine* in July 2003, has been widely discussed. Examples are Gatehouse et al. (2002); Lambrecht (2001); Gerald C. Nelson in Nelson (2001); and Pinstrup-Andersen and Schioler (2000). The original study was Losey et al. (1999). The results of the "remarkable study" were published in the *Proceedings of the National Academy of Sciences USA*, volume 98, issue 21; see especially Sears et al. (2001), which summarizes all the papers, and Stanley-Horn et al. (2001), to which Losey contributed. EPA's 2001 Biopesticides Registration Action Document, "*Bacillus thuringiensis* (Bt) Plant-Incorporated Protectants," which, according to www.scidev.net, shows signs of "a robust social consensus" on the use of Bt corn, is available at www.epa.gov. The taxonomic connection between Bt and anthrax is made by Helgason et al. (2000).

Mellon was quoted by Pringle. Shelton's comment "every entomologist knows" is from Brian A. Federici's chapter in Atherton (2002); "not an entomologist in the world" is from Smith (2000). "How many monarchs get killed on the windshield of a car?" is from "The Pharmageddon Riddle" by Michael Specter, published in the 10 April 2000 *New Yorker*; Specter also designated monarch butterflies "the great fluttering pandas of the insect world." Gordon was quoted in Smith (2000).

The link between plant toxins and natural insecticides is made by Ames and Gold (1997 and 2000); Harlan (1995); and Michael Pariza in Kiple (2002). Nelson's chapter in Nelson (2001) and Smith (2000) discuss the damage done by corn borers; Gianessi (2002) predicts the future use of insecticides in the absence of Bt corn. Chuck Armstrong was interviewed by Fedoroff and Brown in November 2003.

Studies of refugia and insect resistance management, as well as commentaries on the subject, are found in Fox (2003), Ostlie (1997, from which our illustration was adapted), Shelton et al. (2000 and 2002), and Tabashnik et al. (2000 and 2003). Huang et al. (1999) suggests that a dominant form of resistance is possible; in a reply, Shelton and Roush (1999) points out that the resistance Huang saw was unrelated to the presence of Bt in the plant. The *New York Times* reported on compliance with the EPA's refugia rules on 19 June 2003.

11: POLLEN HAS ALWAYS FLOWN

Gene flow, as a natural and ancient process, is discussed in the chapter by Klaus Ammann, Yolande Jacot, Pia Rufener, and Al Mazyad in Custers (2001) and by Ammann (2003).

Papers relevant to transgenes in Mexican maize landraces are Quist and Chapela (2001), the rebuttal by Christou (2002), and the letter by Robert Wager, Peter LaFayette, and Wayne Parrot to the *Electronic Journal of Biotechnology*, 14 August 2002, published on www. agbioworld.org. The dispute was reported in the *New York Times* on 2 October 2001 and 5 April 2002; on the editorial pages of the *Chronicle of Higher Education* (26 April 2002), *Nature* (April 2002 online edition), *Nature Biotechnology* (January 2002), and *Science* (1 March 2002); and by *Berkeleyan* 5 December 2001, *Reason OnLine* 12 February 2002 (reason.com), *Spiked* 17 July 2002 (www.spiked-online.com), and the *Scientist* 6 October 2003 and 10 October 2003.

Martinez-Soriano (2000) places gene flow into the context of corn domestication, as did Major Goodman and Peter Raven in their presentations at the Pew Foundation Conference, Mexico City, September 2003 (pewagbiotech.org/events/0929/presentations/galvez.pdf). Comments by Masa Iwanaga and Julien Berthaud, and the proceedings of the 1995 conference on gene flow in maize, are available at www.cimmyt.cgiar.org. Duvick's remarks on hybrid corn were conveyed by personal correspondence.

Proceedings of the USDA-sponsored conference on canola are McCammon and Dwyer (1990). Hall et al. (2000) reports on canola volunteers with multiple resistance. Rieger et al. (2002) tracked the spread of canola pollen in Australia. Wilkinson et al. (2000 and 2003) report on hybrid brassicas in the U.K.

Bob Scott and Chris Tingle discuss the need for Roundup Ready soy in Arkansas in the *Delta Farm Press*, 7 March 2003.

12: THE ORGANIC RULE

Information on the organic foods industry, including the current Organic Rule, can be found on the USDA's website (www.ams.usda.gov/). Dimitri and Richman (2000) discusses the problems of conflicting standards. A sampling of the controversy following publication of the draft organic rule is available on the website "Pest Management at the Crossroads" (www.pmac.net), including the quotes by Roger Blobaum, Jay Feldman, Dan Glickman, and Eric Kindberg. "Small Organic Farmers Pull Up Stakes" by Samuel Fromartz was published on the Op-Ed page of the *New York Times*, 14 October 2002.

Comparison tests of organic and conventional vegetables were reported in Basker (1992) and Woese et al. (1997). Marian Burros wrote about the Consumers Union study of pesticide residues in the *New York Times*, 8 May 2002; the Associated Press story published the same day was by Philip Brasher.

Baker et al. (2002), Maroni and Fait (1993), and Petrie et al. (2003) discuss pesticide residues and effects. Ames and Gold (1997 and 2000) discuss the contribution of natural and artificial pesticides to cancer, while Block et al. (1992) reports on the epidemiological connection between diets high in fruits and vegetables and cancer. Renn (2003) and Calabrese and Baldwin (2003) examine hormesis. Studies of foodborne bacteria include Bari et al. (2003), Michino et al. (1999), Sagoo et al. (2003), Solomon et al. (2002), and Wachtel et al. (2002). Reports on the possible hazards of Bt include Damgaard et al. (1997), Helgason et al. (2000), and Scribner (2001). Smil (2001) recounts the history of the fertilizer industry.

The study of Washington apples was reported by Tim Steury in *Washington State Magazine*, November 2001. Per Pinstrup-Andersen was interviewed by Brown in February 2003; Pinstrup-Andersen and Schioler (2000) review Altieri's argument concerning the Bolivian potato comparison. Peter Raven's comment that "more is needed" in Africa comes from his address, "The Environmental Challenge," presented at the Natural History Museum in London, 22 May 2003, and posted on www.agbioworld.org.

13: SUSTAINING AGRICULTURE

The successes and failures of the Green Revolution, and the need for new agricultural technologies that conserve land and water resources, are discussed by Avery (1995); Norman Borlaug, in his Thirtieth Anniversary Lecture at the Norwegian Nobel Institute, 8 September 2000 (www.nobel.se); Cassmann (1999 and in a February 2003 interview by Brown); Conway (1997); Evans (1998); Fedoroff and Cohen (1999); Paarlberg (2001); and Tilman (1998). Peter Raven's suggestion that two planets will be needed to sustain a high standard of living comes from "The Environmental Challenge" (www.agbioworld.org).

Andrew C. Revkin wrote "Severe Water and Land Loss Predicted Over a Generation" for the *New York Times*, 23 May 2002. The "water war" in Bolivia was described by William Finnegan in the 8 April 2002 *New Yorker*. Indur Goklany's work was cited by Matt Ridley in the *Guardian*, 3 April 2003. Dennis Avery was interviewed by Jonathan Rauch for the *Atlantic Monthly*, October 2003; the magazine published Gregg Easterbrook's profile of Borlaug in January 1997. The World Resources Institute publication is Thrupp (2002).

Klaus Ammann and Jim Cook were interviewed by Brown in February 2003; Ernie Jaworski and Steve Padgette were interviewed in November 2003. The history of Roundup was provided by Monsanto. David Mortensen of the Pennsylvania State University provided background information on herbicide-resistant soybeans. Ambrosoli (1997) discusses the history of plowing. Information about conservation tillage is available from the Conservation Technology Information Center at Purdue University (www.ctic.purdue.edu), including the report Fawcett and Towery (2002) and a summary of the study of earthworm survival in no-till fields.

14: SHARING THE FRUITS

The story of the San Marzano tomato is told by Michael Specter in the *New Yorker*, 10 April 2000; by Sergio Dompé in *Slow: The International Herald of Taste*, July-September 2001; and by Leonardo Vingiani in the essay, "The Application of Biotechnology: What Do Italians Think?", posted on the website of the International Centre for Pesticides and Health Risk Prevention (www.icps.it). The virus-resistant tomato is reported in Valanzuolo et al. (1994).

Information on plum pox in Pennsylvania comes from a review by Frederick Gildow of the Pennsylvania State University (sharka.cas.psu/review_update.htm); from reports in the USDA's *Agricultural Research*, September 2001 and October 2003 (www.ars.usda.gov); and from the APSNet Feature of March 2000 by Laurene Levy, Vern Damsteegt, Ralph Scorza, and Maria Kolber on the website of the American Phytopathological Society (www.apsnet.org). John Wall interviewed farmers Jim Lerew and Jim Lott for "A Plague Upon the Land," in *Penn State Agriculture*, Summer-Fall 2000. Ravelonandro et al. (2000) describe the virus-resistant plum.

Information on the U.K.'s Farm Scale Evaluations can be found on the website of the Royal Society (www.pubs.royalsoc.ac.uk). The *Scientist* reported on the results on 16 October 2003.

Rob Horsch was interviewed by Brown and Fedoroff in November 2003; some comments are quoted from his 12 June 2003 testimony before the House Science Committee, Subcommittee on Research (www.house.gov/science). Qaim (1999) discusses the economic effects of a virus-resistant sweet potato. Paarlberg (2001) and Conway and Toenniessen (2003) describe Africa's farm production problem. Additional comments come from Robert Paarlberg's lecture, "Environmentally Sustainable Agriculture in the Twenty-first Century," as cited by Horsch in his 12 June 2003 Congressional testimony, and from

Paarlberg's own testimony before the House Science Committee, Sub-committee on Research, 25 September 2001. USAID's Josette Lewis was quoted by Erik Stokstad in the 21 March 2003 *Science*. Other information on the virus-resistant sweet potato came from reprints on Florence Wambugu's website A Harvest Biotech Foundation International (www.ahbfi.org), including a profile by Claire Bisseker in the *Financial Mail*, 24 August 2001, and the interview by the Canadian Broadcasting Corporation, 16 September 2003. Fred Pearce's "Feeding Africa" was published in the *New Scientist*, 27 May 2000. Wambugu's "Taking the Food Out of Our Mouths" was printed in the *Washington Post*, 26 August 2001, and her "Why Africa Needs Agricultural Biotech" was published in *Nature*, 1 July 1999. Lawrence Kent and Nigel Taylor were interviewed by Brown in November 2003.

15: FOOD FOR THOUGHT

Fedoroff and Brown interviewed Eric Sachs and Rob Horsch in November 2003. The *St. Louis Post-Dispatch* editorial appeared 31 October 2003. Brown interviewed Ingo Potrykus in April 2003, John Beard in July 2003, and Karel Schubert in November 2003.

Additional information on the bioavailability tests of IR16844 is posted on the IRRI website (www.irri.org). Robert Russell's calculations were cited by Potrykus in a lecture at Yale University in April 2003. Michael Pollan wrote "Great Yellow Hype" in the *New York Times Magazine*, 4 March 2001. Henry Miller replied to Pollan's article in a letter to the editor submitted to the *New York Times Magazine* on 6 March 2001 and printed on agbioworld.org.

Advances in plant science that could lead to higher yields are described in de la Fuente et al. (1997) on aluminum tolerance; Ku et al. (1999 and 2001) on photosynthesis; Miflin and Habash (2002) on nitrogen use; and Nuccio et al. (1999) and Winicov (1998) on salt stress.

The Zambian decision was reported by Davan Maharaj and Anthony Mukwita for the *Los Angeles Times* (28 August 2002) and by Sudarsan Raghavan for Knight-Ridder newspapers (24 November 2002). Zimbabwe's situation was reported by the BBC on 6 September 2002. The *New York Times* article, in which Lizette Alvarez reported from the Happy Apple greengrocer in Totnes, was published 11 February 2003. Dan Simberloff was quoted in the *Washington Post* 21 April 1998. Fussell (1992) identifies the "world of hidden corn." Simon Mwanza describes his trip to America, during which he learned that "Maize is not directly consumed in America," in the *Times of Zambia* (26 June 2003); his article is available on allafrica.com and www.checkbiotech.org.

The 2003 survey in America was performed by the Discovery Channel and reported by the International Service for the Acquisition of Agri-biotech Applications (www.isaaa.org). The 1993 survey of New Jersey residents was performed by Rutgers University and the New Jersey Agricultural Experiment Station and is available through the National Agriculture Library at www.nal.usda.gov/bic/Pubpercep/.

The need for more food, and the environmental challenge of producing it, are discussed by Norman Borlaug in an essay printed on the *New York Times* Op-Ed page, 11 July 2003; and by Avery (1995), Brown (1995), Cassman (1999), Cohen (1995), Conway and Toenniessen (2003), Ehrlich (1968), Hardin (1968), Miller (2003), and Smil (2001). Gary Toenniessen was interviewed by Brown in February 2003. Juma (2003) addresses the ideologies that stand in the way of accepting genetically modified foods. Peter Raven's comments are from his address, "The Environmental Challenge" (www.agbioworld.org).

BIBLIOGRAPHY

Allen, G. E. 1975. Life Science in the Twentieth Century. New York: John Wiley and Sons.

Ambrosoli, M. 1997. The Wild and the Sown: Botany and Agriculture in Western Europe, 1350-1850. Cambridge: Cambridge University Press.

Ames, B. N. and L. S. Gold. 1997. Environmental pollution, pesticides, and the prevention of cancer: misconceptions. FASEB Journal 11:1041-52.

Ames, B. N. and L. S. Gold. 2000. Paracelsus to parascience: the environmental cancer distraction. Mutation Research 447:3-13.

Ammann, K. 2003. Biodiversity and Agricultural Biotechnology: A Review of the Impact of Agricultural Biotechnology on Biodiversity. Bern, Switzerland: University of Bern Botanical Garden.

Anderson, E. 1967. Plants, Man, and Life. Berkeley: University of California Press.

Anderson, P. C., ed. 1999. The Prehistory of Agriculture. Los Angeles: University of California Press.

Atherton, K. T., ed. 2002. Genetically Modified Crops: Assessing Safety. London and New York: Taylor and Francis.

Avery, D. 1995. Saving the Planet with Pesticides and Plastic: The Environmental Triumph of High-Yield Farming. Indianapolis: Hudson Institute.

Baker, B. P., C. M. Benbrook, E. Groth 3rd, and K. Lutz Benbrook. 2002. Pesticide residues in conventional, integrated pest management (IPM)-grown and organic foods: insights from three U.S. data sets. Food Additives and Contaminants 19:427-46.

Bari, M. L., E. Nazuka, Y. Sabina, S. Todoriki, and K. Isshiki. 2003. Chemical and irradiation treatments for killing Escherichia coli O157:H7 on alfalfa, radish, and mung bean seeds. Journal of Food Protection 66:767-74.

Basker, D. 1992. Comparison of taste quality between organically and conventionally grown fruits and vegetables. American Journal of Alternative Agriculture 7:129-36.

Battelli, M. G., L. Barbieri, A. Bolognesi, L. Buonamici, P. Valbonesi, L. Polito, E. J. Van Damme, W. J. Peuman, and F. Stirpe. 1997. Ribosome-inactivating lectins with polynucleotide: adenosine glycosidase activity. FEBS Letters 408:355-59.

Benes, P., ed. 1996. Plants and People: The Dublin Seminar for New England Folklife Annual Proceedings 1995. Boston: Boston University Press.

Bennetzen, J., E. Buckler, V. Chandler, J. Doebley, J. Dorweiler, B. Gaut, M. Freeling, S. Hake, E. Kellogg, R. S. Poethig, V. Walbot, and S. Wessler. 2001. Genetic evidence and the origin of maize. Latin American Antiquity 12:84-86.

Benz, B. F. 2001. Archaeological evidence of teosinte domestication from Guila Naquitz, Oaxaca. Proceedings of the National Academy of Sciences USA 98:2104-6.

Berg, P., D. Baltimore, H. W. Boyer, S. N. Cohen, R. W. Davis, D. S. Hogness, D. Nathans, R. Roblin, J. D. Watson, S. Weissman, and N. D. Zinder. 1974. Potential biohazards of recombinant DNA molecules. Science 185:303.

Betz, F. S., B. G. Hammond, and R. L. Fuchs. 2000. Safety and advantages of *Bacillus thuringiensis*-protected plants to control insect pests. Regulatory Toxicology and Pharmacology 32:156-73.

Beyer, P., S. Al-Babili, X. Ye, P. Lucca, P. Schaub, R. Welsch, and I. Potrykus. 2002. Golden Rice: introducing the beta-carotene biosynthesis pathway into rice endosperm by genetic engineering to defeat vitamin A deficiency. Journal of Nutrition 132:506S-510S.

Bhat, S. F., and S. Srinivasan. 2002. Molecular and genetic analyses of transgenic plants: considerations and approaches. Plant Science 163:673-81.

Bhattacharyya, M. K., A. M. Smith, T. H. Ellis, C. Hedley, and C. Martin. 1990. The wrinkled-seed character of pea described by Mendel is caused by a transposon-like insertion in a gene encoding starch-branching enzyme. Cell 60:115-22.

Block, G., B. Patterson, and A. Subar. 1992. Fruit, vegetables, and cancer prevention: a review of the epidemiological evidence. Nutrition and Cancer 18:1-29.

Brown, J. R. 2003. Ancient horizontal gene transfer. Nature Reviews Genetics 4:121-32.

Brown, L. R. 1995. Who Will Feed China? Wake-Up Call for a Small Planet. New York: W. W. Norton.

Buchanan, B. 2001. Genetic engineering and the allergy issue. Plant Physiology 126:5-7.

Buckler, E. S. T., J. M. Thornsberry, and S. Kresovich. 2001. Molecular diversity, structure and domestication of grasses. Genetic Research 77:213-218.

Bud, R. 1993. The Uses of Life: A History of Biotechnology. Cambridg: Cambridge University Press.

Busch, L., W. B. Lacy, J. Burkhardt, and L. R. Lacy. 1991. Plants, Power, and Profit: Social, Economic, and Ethical Consequences of the New Biotechnologies. Cambridge and Oxford: Blackwell.

Calabrese, E. J., and L. A. Baldwin. 2003. Hormesis: the dose-response revolution. Annual Reviews of Pharmacology and Toxicology 43:175-97.

Carlini, C. R., and M. F. Grossi-de-Sa. 2002. Plant toxic proteins with insecticidal properties: a review of their potentialities as bioinsecticides. Toxicon 40:1515-39.

Carlson, E. A. 1966. The Gene: A Critical History. Philadelphia: W. B. Saunders.

Cassman, K. G. 1999. Ecological intensification of cereal production systems: yield potential, soil quality, and precision agriculture. Proceedings of the National Academy of Sciences USA 96:5952-59.

Chambers, P. A., P. S. Duggan, J. Heritage, and J. M. Forbes. 2002. The fate of antibiotic resistance marker genes in transgenic plant feed material fed to chickens. Journal of Antimicrobial Chemotherapy 49:161-64.

Cheng, C.-H. C. 1998. Evolution of the diverse antifreeze proteins. Current Opinions in Genetics and Development 8:715-20.

Chilton, M-D., M. H. Drummond, D. J. Merio, D. Sciaky, A. L. Montoya, M. P. Gordon, and E. W. Nester. 1977. Stable incorporation of plasmid DNA into higher plant cells: the molecular basis of crown gall tumorgenesis. Cell 11:263-71.

Christou, P. 2002. No credible scientific evidence is presented to support claims that transgenic DNA was introgressed into traditional maize landraces in Oaxaca, Mexico. Transgenic Research 11:iii-v.

Coe, S. D., and M. D. Coe. 1996. The True History of Chocolate. London: Thames and Hudson.

Cohen, J. E. 1995. How Many People Can the Earth Support? New York: Norton.

Cohen, M. N. 1997. The Food Crisis in Prehistory: Overpopulation and the Origins of Agriculture. New Haven: Yale University Press.

Cohen, S. N., A. C. Chang, H. W. Boyer, and R. B. Helling. 1973. Construction of biologically functional bacterial plasmids in vitro. Proceedings of the National Academy of Sciences USA 70:3240-44.

Conway, G. 1997. The Doubly Green Revolution: Food for All in the Twenty-first Century. Ithaca: Cornell University Press.

Conway, G., and G. Toenniessen. 2003. Science for African food security. Science 299:1187-88.

Cowan, C. W., and P. J. Watson, eds. 1992. The Origins of Agriculture: An International Perspective. Washington, D.C.: Smithsonian Institution Press.

Custers, R., ed. 2001. Safety of Genetically Engineered Crops. Flanders: Inter-university Institute for Biotechnology, VIB publication.

Damgaard, P. H., P. E. Granum, J. Bresciani, M. V. Torregrowwa, J. Eilenberg, and L. Valentino. 1997. Characterization of *Bacillus thuringiensis* isolated from infections in burn wounds. FEMS Immunology and Medical Microbiology 18:47-53.

Davis, B. D., ed. 1991. The Genetic Revolution: Scientific Prospects and Public Perceptions. Baltimore and London: Johns Hopkins University Press.

Diamond, J. 2002. Evolution, consequences, and future of plant and animal domestication. Nature 418:700-707.

Diamond, J. 1997; rev. 1999. Guns, Germs, and Steel. New York: W. W. Norton.

Dimitri, C., and N. J. Richman. 2000. Organic Food Markets in Transition: Policy Studies Reports, ISSN 1521-8201. Greenbelt, Md.: Henry A. Wallace Center for Agricultural & Environmental Policy.

Doebley, J. 1994. Morphology, molecules, and maize. In Corn and Culture in the Prehistoric New World, S. Johannessen and C. A. Hastorf, eds. Boulder, Co.: Westview Press.

Doebley, J. 1992. Mapping the genes that made maize. Trends in Genetics 8:302-7.

Doebley, J. 1997. A brief note on the rediscovery of Durango teosinte. Maize Genetics Cooperation Newsletter 57. Online. Available at www.agron.missouri.edu/mnl/57/97doebley.html.

Doebley, J. 2001. George Beadle's other hypothesis: one-gene, one-trait. Genetics 158:487-93.

Doebley, J., A. Stec, and L. Hubbard. 1997. The evolution of apical dominance in maize. Nature 386:485-88.

Doerfler, W. 2000. Foreign DNA in Mammalian Systems. Weinheim, Germany: Wiley-VCH.

Down, R. E., L. Ford, S. D. Woodhouse, R. J. Raemaekers, B. Leitch, J. A. Gatehouse, and A. M. Gatehouse. 2000. Snowdrop lectin (GNA) has no acute toxic effects on a beneficial insect predator, the 2-spot ladybird (Adalia bipunctata L.). Journal of Insect Physiology 46:379-91.

Down, R. E., L. Ford, S. J. Bedford, L. N. Gatehouse, C. Newell, J. A. Gatehouse, and A. M. Gatehouse. 2001. Influence of plant development and environment on transgene expression in potato and consequences for insect resistance. Transgenic Research 10:223-36.

Dreyer, P. 1985. A Gardener Touched with Genius: The Life of Luther Burbank. Los Angeles: University of California Press.

Duboule, D., and A. S. Wilkins. 1998. The evolution of "bricolage." Trends in Genetics 14:54-59.

Duggan, P. S., P.A. Chambers, J. Heritage, and J. M. Forbes. 2000. Survival of free DNA encoding antibiotic resistance from transgenic maize and the transformation activity of DNA in ovine saliva, ovine rumen fluid, and silage effluent. FEMS Microbiology Letters 191:71.

Ehrlich, P. 1968. The Population Bomb. New York: Ballantine.

Evans, L. T. 1998. Feeding the Ten Billion: Plants and Population Growth. Cambridge: Cambridge University Press.

Ewen, S. W., and A. Pusztai. 1999. Effect of diets containing genetically modified potatoes expressing Galanthus nivalis lectin on rat small intestine. Lancet 354:1353-54.

Eyre-Walker, A., R. L. Gaut, H. Hilton, D. L. Feldman, and B. S. Gaut. 1998. Investigation of the bottleneck leading to the domestication of maize. Proceedings of the National Academy of Sciences USA 95:4441-46.

Farb, P., and G. Armelagos. 1980. Consuming Passions: The Anthropology of Eating. Boston: Houghton Mifflin.

Fawcett, R., and D. Towery. 2002. Conservation Tillage and Plant Biotechnology: How New Technologies Can Improve the Environment by Reducing the Need to Plow. W. Lafayette, Ind.: Conservation Technology Information Center.

Fedoroff, N. V., and J. E. Cohen. 1999. Plants and population: Is there time? Proceedings of the National Academy of Sciences USA 96:5903-07.

Fernandez-Armesto, F. 2002. Near a Thousand Tables: A History of Food. New York: Free Press.

Fitches, E., S. D. Woodhouse, J. P. Edwards, and J. A. Gatehouse. 2001. In vitro and in vivo binding of snowdrop (Galanthus nivalis agglutinin; GNA) and jackbean (Canavalia ensiformis; Con A) lectins within tomato moth (Lacanobia oleracea) larvae; mechanisms of insecticidal action. Journal of Insect Physiology 47:777-87.

Fitches, E., N. Audsley, J. A. Gatehouse, and J. P. Edwards. 2002. Fusion proteins containing neuropeptides as novel insect contol agents: snowdrop lectin delivers fused allatostatin to insect haemolymph following oral ingestion. Insect Biochemistry and Molecular Biology 32:1653-61.

Foissac, X., N. Thi Loc, P. Christou, A. M. Gatehouse, and J. A. Gatehouse. 2000. Resistance to green leafhopper (*Nephotettix virescens*) and brown planthopper (*Nilaparvata lugens*) in transgenic rice expressing snowdrop lectin (*Galanthus nivalis* agglutinin; GNA). Journal of Insect Physiology 46:573-83.

Food and Agriculture Organization of the United Nations and World Health Organization (FAO-WHO). 2000. Safety Aspects of Genetically Modified Foods of Plant Origin: Report of a Joint FAO-WHO Expert Consultation on Foods Derived from Biotechnology. Geneva: World Health Organization.

Food and Agriculture Organization of the United Nations and World Health Organization (FAO-WHO). 2001. Safety Assessments of Foods Derived From Genetically Modified Microorganisms: Report of a Joint FAO-WHO Expert Consultation on Foods Derived from Biotechnology. Geneva: World Health Organization.

Food and Agriculture Organization of the United Nations and World Health Organization (FAO-WHO). 2001. Evaluation of Allergenicity of Genetically Modified Foods: Report of a Joint FAO-WHO Expert Consultation on Allergenicity of Foods Derived from Biotechnology. Rome: Food and Agriculture Organization of the United Nations.

Fox, J. L. 2003. Resistance to Bt toxin surprisingly absent in pests. Nature Biotechnology 21:958-59.

Freed, D. 1999. Do dietary lectins cause disease? British Medical Journal 318:1023-4.

Fuente, J. M. de la, V. Ramirez-Rodriguez, J. L. Cabrera-Ponce, and L. Herrera-Estrella. 1997. Aluminum tolerance in transgenic plants by alteration of citrate synthesis. Science 276:1566-68.

Fussell, B. 1992. The Story of Corn. New York: Knopf.

Gatehouse, A. M. R., N. Ferry, and R. J. M. Raemaekers. 2002. The case of the monarch butterfly: a verdict is returned. Trends in Genetics 18:249-51.

Gianessi, L. P., C. S. Silvers, S. Sankula, and J. E. Carpenter. 2002. Plant Biotechnology: Current and Potential Impact for Improving Pest Management in U.S. Agriculture, an Analysis of 40 Case Studies. Washington, D.C.: National Center for Food and Agricultural Policy.

Hall, L., K. Topinka, J. Huffmann, L. Davis, and A. Good. 2000. Pollen flow between herbicide-resistant *Brassica napus* is the cause of multiple-resistant *B. napus* volunteers. Weed Science 48:688–94.

Hallman, W. K., and J. Metcalfe. 1993. Public perceptions of agricultural biotechnology: a survey of New Jersey residents. New Brunswick, N.J.: Ecosystem Policy Research Center, Rutgers University, and New Jersey Agricultural Experiment Station, Cook College.

Hardin, G. 1968. The tragedy of the commons. Science 162:1243-48.

Harlan, J. R. 1995. The Living Fields. Cambridge: Cambridge University Press.

Harris, D. R., and G. C. Hillman, eds. 1989. Foraging and Farming: The Evolution of Plant Exploitation. London: Unwin Hyman.

Harten, A. M. van. 1998. Mutation Breeding: Theory and Practical Applications. Cambridge: Cambridge University Press.

Heiser, C. B., Jr. 1990. Seed to Civilization: The Story of Food. 2nd ed. rev. Cambridge: Harvard University Press.

Helgason, E., O. A. Okstad, D. A. Caugant, H. A. Johansen, A. Fouet, M. Mock, I. Hegna, and A. B. Kolstø. 2000. *Bacillus anthracis, Bacillus cereus,* and *Bacillus thuringiensis*—one species on the basis of genetic evidence. Applied Environmental Microbiology 66:2627-30.

Hester, G., and C. S. Wright. 1996. The mannose-specific bulb lectin from *Galanthus nivalis* (snowdrop) binds mono- and dimannosides at distinct sites: structure analysis of refined complexes at 2.3 A and 3.0 A resolution. Journal of Molecular Biology 262:516-31.

Hey, J. 1997. A reduction of "species" resolves the species problem. Online. Available at http://lifesci.rutgers.edu/~heylab/sconcept/introduction.html.

Hightower, R., C. Baden, E. Penzes, P. Lund, and P. Dunsmuir. 1991. Expression of antifreeze proteins in transgenic plants. Plant Molecular Biology 17:1013-21.

Ho, M.-W., A. Ryan, and J. Cummins. 1999. Cauliflower mosaic viral promoter—a recipe for disaster. Microbial Ecology in Health and Disease 11:194.

Huang, F., L. L. Buschman, R. A. Higgins, and W. H. McGaughey. 1999. Inheritance of resistance to *Bacillus thuringiensis* toxin (Dipel ES) in the European corn borer. Science 284:965-67.

Huang, S., A. Sirikhachornkit, X. Su, J. Faris, B. Gill, R. Haselkorn, and P. Gornicki. 2002. Genes encoding plastid acetyl-CoA carboxylase and 3-phosphoglycerate kinase of the *Triticum/Aegilops* complex and the evolutionary history of polyploid wheat. Proceedings of the National Academy of Sciences USA 99:8133-38.

Huang, T., J. Nicodemus, D. G. Zarka, M. F. Thomashow, M. Wisniewski, and J. G. Duman. 2002. Expression of an insect (*Dendroides canadensis*) antifreeze protein in *Arabidopsis thaliana* results in a decrease in plant freezing temperature. Plant Molecular Biology 50:333-44.

Hubbell, S. 2001. Shrinking the Cat: Genetic Engineering Before We Knew About Genes. New York: Houghton Mifflin.

Ichikawa, T., M. Nakazawa, M. Kawashima, S. Muto, K. Gohda, K. Suzuki, A. Ishikawa, H. Kobayashi, T. Yoshizumi, Y. Tsumoto, Y. Tsuhara, H. Iizumi, Y. Goto, and M. Matsui. 2003. Sequence database of 1172 T-DNA insertion sites in *Arabidopsis* activation-tagging lines that showed phenotypes in T1 generation. Plant Journal 36:421-29.

Iltis, H. 1983. From teosinte to maize: the catastrophic sexual transmutation. Science 222:886-94.

Jackson, D. A., R. H. Symons, and P. Berg. 1972. Biochemical method for inserting new genetic information into DNA of Simian Virus 40: circular SV40 DNA molecules containing lambda phage genes and the galactose operon of *Escherichia coli.* Proceedings of the National Academy of Sciences USA 69:2904-9.

Jacob, F. 1977. Evolution and tinkering. Science 196:1161-66.

Jaenicke-Despres, V., E. S. Buckler, B. D. Smith, M. T. P. Gilbert, A. Cooper, J. Doebley, and S. Pääbo. 2003. Early allelic selection in maize as revealed by ancient DNA. Science 302:1206-08.

Jain, S. Mohan, D. S. Brar, and B. S. Ahloowalia, eds. 1998. Somaclonal Variation and Induced Mutations in Crop Improvement. Dordrecht, Boston, and London: Kluwer Academic Publishers.

Jain, S. Mohan, D. S. Brar, and B. S. Ahloowalia, eds. 2002. Molecular Techniques in Crop Improvement. Dordrecht, Boston, and London: Kluwer Academic Publishers.

Jones, R. 1969. Fire-stick farming. Australian Natural History 167:224-28.

Jones, R. 1995. Mindjongork: legacy of the firestick. In Country in Flames; Proceedings of the 1994 symposium on biodiversity and fire in North Australia, D. Rose, ed. Canberra & Darwin: Biodiversity Unit, Department of the Environment, Sport and Territories, and the North Australia Research Unit.

Judson, H. F. 1979. The Eighth Day of Creation: Makers of the Revolution in Biology. New York: Simon and Schuster.

Juma, C. 2003. Biotechnology in the global communication ecology. Economic Perspectives. Online. Available at http://usinfo.state.gov/journals/ites/0903/ijee/juma.htm.

Kashkush, K., M. Feldman, and A. A. Levy. 2002. Gene loss, silencing, and activation in a newly synthesized wheat allotetraploid. Genetics 160:1651-59.

Kass, L. B., and C. Bonneuil. 2004. Mapping and seeing: Barbara McClintock and the linking of genetics and cytology in maize genetics, 1928-1935. In Classic Genetic Research and its Legacy: The Mapping Cultures of Twentieth Century Genetics, J.-P. Gaudilliere and H.-J. Rheinberger, eds. London: Routledge.

Kellogg, E. A. 2001. Evolutionary history of the grasses. Plant Physiology 125:1198-1205.

Kelman, A., W. Anderson, S. Falkow, N. Fedoroff, and S. Levin. 1987. Introduction of Recombinant DNA-Engineered Organisms into the Environment: Key Issues. Washington, D.C.: National Academy Press.

Khush, G. S. 1997. Origin, dispersal, cultivation and variation of rice. Plant Molecular Biology 35:25-34.

Kiesselbach, T. A. 1999. The structure and reproduction of corn: 50th anniversary edition. New York: Cold Spring Harbor Lab.

Kiple, K., ed. 2002. The Cambridge World History of Food. Cambridge: Cambridge University Press.

Kleter, G., and A. Peijnenburg. 2002. Screening of transgenic proteins expressed in transgenic food crops for the presence of short amino acid sequences identical to potential, IgE-binding linear epitopes of allergens. BMC Structural Biology 2:8-19.

Kloppenburg, J. R., Jr. 1988. First the Seed: The Political Economy of Plant Biotechnology, 1492-2000. Cambridge: Cambridge University Press.

Kohli A., S. Griffiths, N. Palacios, R. M. Twyman, P. Vain, D. A. Laurie, and P. Christou. 1999. Molecular characterization of transforming plasmid rearrangements in transgenic rice reveals a recombination hotspot in the CaMV 35S promoter and confirms the predominance of microhomology mediated recombination. Plant Journal 17:591-601.

Kozukue, N., S. Misoo, T. Yamada, O. Kamijima, and M. Friedman. 1999. Inheritance of morphological characters and glycoalkaloids in potatoes of somatic hybrids between dihaploid *Solanum acaule* and tetraploid *Solanum tuberosum*. Journal of Agricultural and Food Chemistry 47:4478-83.

Ku, M. S., S. Agarie, M. Nomura, H. Fukayama, H. Tsuchida, K. Ono, S. Hirose, S. Toki, M. Miyao, and M. Matsuoka. 1999. High-level expression of maize phosphoenolpyruvate carboxylase in transgenic rice plants. Nature Biotechnology 17:76-80.

Ku, M. S., D. Cho, X. Li, D. M. Jiao, M. Pinto, M. Miyao, and M. Matsuoka. 2001. Introduction of genes encoding C4 photosynthesis enzymes into rice plants: physiological consequences. Novartis Foundation Symposium 236:100-111.

Lambrecht, B. 2001. Dinner at the New Gene Café: How Genetic Engineering is Changing What We Eat, How We Live, and the Global Politics of Food. New York: St. Martin's Press.

Lauter, N., and Doebley, J. 2002. Genetic variation for phenotypically invariant traits detected in teosinte: implications for the evolution of novel forms. Genetics 160:333-42.

Losey, J. E., L. S. Rayor, and M. E. Carter. 1999. Transgenic pollen harms monarch larvae. Nature 399:214.

Lucca, P., X. Ye, and I. Potrykus. 2001. Effective selection and regeneration of transgenic rice plants with mannose as selective agent. Molecular Breeding 7:43-49.

Lurquin, P. 2001. The Green Phoenix: A History of Genetically Modified Plants. New York: Columbia University Press.

Lurquin, P. 2002. High Tech Harvest: Understanding Genetically Modified Food Plants. Boulder and Oxford: Westview Press.

MacNeish, R. S. 1991. The Origins of Agriculture and Settled Life. Norman and London: University of Oklahoma Press.

Magnus, D., A. Caplan, and G. McGee, eds. 2002. Who Owns Life? Amherst: Prometheus Books.

Maluszynski, M., and K. J. Kasha, eds. 2002. Mutations, In Vitro and Molecular Techniques for Environmentally Sustainable Crop Improvement. Dordrecht, Boston, and London: Kluwer Academic Publishers.

Maroni, M., and A. Fait. 1993. Health effects in man from long-term exposure to pesticides: a review of the 1975-1991 literature. Toxicology 78:1-180.

Martineau, B. 2001. First Fruit: The Creation of the FlavrSavr Tomato and the Birth of Genetically Engineered Food. New York: McGraw-Hill.

Martinez-Soriano, J. P. R. 2000. Transgenic maize in Mexico: no need for concern. Science 287:1399.

Matsuoka, Y., Y. Vigouroux, M. M. Goodman, G. J. Sanchez, E. Buckler, and J. Doebley. 2002. A single domestication for maize shown by multilocus microsatellite genotyping. Proceedings of the National Academy of Sciences USA 99:6080-84.

McCammon, S. L., and S. G. Dwyer, eds. 1990. Workshop on Safeguards for Planned Introduction of Transgenic Oilseed Crucifers: Proceedings. Washington, D.C.: U.S. Department of Agriculture.

McHughen, A. 2002. Pandora's Picnic Basket: The Potential and Hazards of Genetically Modified Foods. Oxford, U.K.: Oxford University Press.

Michino, H., K. Araki, S. Minami, S. Takaya, N. Sakai, M. Miyazaki, A. Ono, and H. Yanagawa. 1999. Massive outbreak of *Escherichia coli* O157:H7 infection in schoolchildren in Sakai City, Japan, associated with consumption of white radish sprouts. American Journal of Epidemiology 150:787-96.

Miflin, B. J., and D. Z. Habash. 2002. The role of glutamine synthetase and glutamate degydrogenase in nitrogen assimilation and possibilities for improvement in the nitrogen utilization of crops. Journal of Experimental Botany 53:979-87.

Miller, H. 2003. Vox populi and public policy: why should we care? Nature Biotechnology 21:1431-32.

Morrison, L. A. 1998. Taxonomic issues in *Triticum* L. and *Aegilops* L. Wheat Information Service 86:49-53.

Mullis, K. 1998. Dancing Naked in the Mind Field. New York: Pantheon Books.

Naughton P. J., G. Grant, S. Bardocz, and A. Pusztai. 2000. Modulation of *Salmonella* infection by the lectins of *Canavalia ensiformis* (Con A) and *Galanthus nivalis* (GNA) in a rat model in vivo. Journal of Applied Microbiology 88:720-27.

Nelson, G. C., ed. 2001. Genetically Modified Organisms in Agriculture: Economics and Politics. San Diego and London: Academic Press.

Nestle, M. 1996. Allergies to transgenic foods—questions of policy. The New England Journal of Medicine 334:726-28.

Netherwood, T., S. M. Martín-Orúe, A. G. O'Donnell, S. Gockling, J. Graham, J. C. Mathers, and H. J. Gilbert. 2004. Assessing the survival of transgenic plant DNA in the human gastrointestinal tract. Nature Biotechnology 22:204-9.

Nordlee, J. A., S. L. Taylor, J. A. Townsend, L. A. Thomas, and R. K. Bush. 1996. Identification of a Brazil-nut allergen in transgenic soybeans. New England Journal of Medicine 334:688-92.

Nuccio, M. L., D. Rhodes, S. D. McNeil, and A. D. Hanson. 1999. Metabolic engineering of plants for osmotic stress resistance. Current Opinions in Plant Biology 2:128-34.

Oka, H. I. 1988. Origin of Cultivated Rice. Japan Scientific Societies Press Developments in Crop Science 14. Tokyo: Elsevier.

Olsnes, S., and J. V. Kozlov. 2001. Ricin. Toxicon 39:1723-28.

Orlean, S. 1998. The Orchid Thief. New York: Random House.

Ostlie, K. R., W. D. Hutchison, and R. L. Hellmich. 1997. Bt corn and European corn borer. NCR publication 602. St. Paul: University of Minnesota.

Paarlberg, R. L. 2001. The Politics of Precaution: Genetically Modified Crops in Developing Countries. Baltimore and London: International Food Policy Research Institute.

Patel B., R. Schutte, P. Sporns, J. Doyle, L. Jewel, and R. N. Fedorak. 2002. Potato glycoalkaloids adversely affect intestinal permeability and aggravate inflammatory bowel disease. Inflammatory Bowel Diseases 8:340-46.

Pendergast, M. 1999. Uncommon Grounds. New York: Basic Books.

Petrie, K., M. Thomas, and E. Broadbent. 2003. Symptom complaints following aerial spraying with biological insecticide Foray 48B. New Zealand Medical Journal 116:U354

Pinstrup-Andersen, P., and E. Schioler. 2000. Seeds of Contention: World Hunger and the Global Controversy Over GM Crops. Baltimore and London: International Food Policy Research Institute.

Piperno, D. R., and K. V. Flannery. 2001. The earliest archaeological maize *Zea mays* L. from highland Mexico: new accelerator mass spectrometry dates and their implications. Proceedings of the National Academy of Sciences USA 98:2101-03.

Pollan, M. 2001. The Botany of Desire: A Plant's-Eye View of the World. New York: Random House.

Portugal, F. H., and J. S. Cohen. 1977. A Century of DNA. Cambridge: MIT Press.

Potrykus, I. 2001. Golden Rice and beyond. Plant Physiology 125:1157-61.

Pusztai A., S. W. Ewen, G. Grant, W. J. Peumans, E. J. van Damme, L. Rubio, and S. Bardocz. 1990. Relationship between survival and binding of plant lectins during small intestinal passage and their effectiveness as growth factors. Digestion 46 Suppl 2: 308-16.

Pusztai A., G. Grant, R. J. Spencer, T. J. Duguid, D. S. Brown, S. W. Ewen, W. J. Peumans, E. J. van Damme, and S. Bardocz. 1993. Kidney bean lectin-induced *Escherichia coli* overgrowth in the small intestine is blocked by GNA, a mannose-specific lectin. Journal of Applied Bacteriology 75:360-68.

Qaim, M. 1999. The Economic Effects of Genetically Modified Orphan Commodities: Projections for Sweetpotato in Kenya. ISAAA Briefs No. 13. Ithaca, N.Y., and Bonn: ISAAA and Center for Development Research (ZEF).

Quist, D., and I. Chapela. 2001. Transgenic DNA introgressed into traditional maize landraces in Oaxaca, Mexico. Nature 414:541-43.

Rabinow, P. 1996. Making PCR: A Story of Biotechnology. Chicago and London: University of Chicago Press.

Rao, K. V., K. S. Rathore, T. K. Hodges, X. Fu, E. Stoger, D. Sudhakar, S. Williams, P. Christou, M. Bharathi, D. P. Bown, K. S. Powell, J. Spence, A. M. Gatehouse, and J. A. Gatehouse. 1998. Expression of snowdrop lectin (GNA) in transgenic rice plants confers resistance to rice brown planthopper. Plant Journal 15:469-77.

Ravelonandro, M., R. Scorza, A. Callahan, L. Levy, C. Jácquet, M. Monsion, and V. Damsteegt. 2000. The use of transgenic fruit trees as a resistance strategy for virus epidemics: the plum pox (sharka) model. Virus Research 71:63-69.

Rayle, D., and L. Wedberg. 1975. Botany: A Human Concern. New York: Houghton Mifflin.

Renn, O. 2003. Hormesis and risk communication. Human and Experimental Toxicology 22:3-24.

Rieger, M. A., M. Lamond, C. Preston, S. B. Powles, and R. T. Roush. 2002. Pollen-mediated movement of herbicide resistance between commercial canola fields. Science 296:2386-88.

Roberts, J. 2001. The Origins of Fruits and Vegetables. New York: Universe Publishing.

Roelofs, J., and P. J. Van Haastert. 2001. Genes lost during evolution. Nature 411:1013-14.

Romeis, J., D. Babendreier, and F. L. Wackers. 2003. Consumption of snowdrop lectin (*Galanthus nivalis* agglutinin) causes direct effects on adult parasitic wasps. Oecologia 134:528-36.

Rudiger H., H. C. Siebert, D. Solis, J. Jimenez-Barbero, A. Romero, C. W. von der Lieth, T. Diaz-Marino, and H. J. Gabius. 2000. Medicinal chemistry based on the sugar code: fundamentals of lectinology and experimental strategies with lectins as targets. Current Medicinal Chemistry 7:389-416.

Russo, E. 2003. The birth of biotechnology. Nature 421:456-57.

Sagoo, S. K., C. L. Little, L. Ward, I. A. Gillespie, and R. T. Mitchell. 2003. Microbiological study of ready-to-eat salad vegetables from retail establishments uncovers a national outbreak of salmonellosis. Journal of Food Protection 66:403-9.

Salamini, F., F. H. Ozkan, A. Brandolini, R. Schafer-Pregl, and W. Martin. 2002. Genetics and geography of wild cereal domestication in the near east. Nature Reviews Genetics 3:429-41.

Salvador, R. J. 1998. Maize. In The Encyclopedia of Mexico: History, Culture, and Society, M. Werner, ed. New York: Fitzroy Dearborn.

Sanford, J. C. 1988. The biolistic process. Trends in Biotechnology 6:229-302.

Sanford, J. C. 2000. The development of the biolistic process. In Vitro Cellular and Developmental Biology 36:303-8.

Sasaki, A., M. Ashikari, M. Ueguchi-Tanaka, H. Itoh, A. Nishimura, D. Swapan, K. Ishiyama, T. Saito, M. Kobayashi, G. S. Khush, H. Kitano, and M. Matsuoka. 2002. Green Revolution: a mutant gibberellin-synthesis gene in rice. Nature 416:701-2.

Schell, J., and M. Van Montagu. 1977. Transfer, maintenance, and expression of bacterial Ti-plasmid DNA in plant cells transformed with *A. tumefaciens*. Brookhaven Symposia Biology 29:36-49.

Schubbert, R., D. Renz, B. Schmitz, and W. Doerfler. 1997. Foreign (M13) DNA ingested by mice reaches peripheral leukocytes, spleen, and liver via the intestinal wall mucosa and can be covalently linked to mouse DNA. Proceedings of the National Academy of Sciences USA 94:961-66.

Schubbert, R., U. Hohlweg, D. Renz, and W. Doerfler. 1998. On the fate of orally ingested foreign DNA in mice: chromosomal association and placental transfer to the fetus. Molecular and General Genetics 259:569-76.

Scribner, M. J. 2001. Bt or not Bt: is that the question? Proceedings of the National Academy of Sciences USA 98:12328-30.

Sears, M. K., R. L. Hellmich, D. E. Stanley-Horn, K. S. Oberhauser, J. M. Pleasant, H. R. Mattila, B. D. Siegfried, and G. P. Dively. 2001. Impact of Bt corn pollen on monarch butterfly populations: a risk assessment. Proceedings of the National Academy of Sciences USA 98:11937-42.

Shelton, A. M., and R. T. Roush. 1999. False reports and the ears of men. Nature Biotechnology 17:832.

Shelton, A. M., J. D. Tang, R. T. Roush, T. D. Metz, and E. D. Earle. 2000. Field tests on managing resistance to Bt-engineered plants. Nature Biotechnology 18:339-42.

Shine, I., and S. Wrobel. 1976. Thomas Hunt Morgan: Pioneer of Genetics. Lexington: University Press of Kentucky.

Shouse, B. 2001. Spreading the word, scattering the seeds. Science 294:988-89.

Siegel, J. P. 2001. The mammalian safety of *Bacillus thuringiensis*-based insecticides. Journal of Invertebrate Pathology 77:13-21.

Simmonds, N. W., ed. 1976. The Evolution of Crop Plants. London and New York: Longman.

Simpson, B. B., and M. C. Ogorzaly. 1995. Economic Botany: Plants in Our World. Rev. ed. New York: McGraw Hill.

Smil, V. 2001. Enriching the Earth: Fritz Haber, Carl Bosch, and the Transformation of World Food Production. Cambridge: MIT Press.

Smith, B. D. 2001. Documenting plant domestication: the consilience of biological and archaeological approaches. Proceedings of the National Academy of Sciences USA 98:1324-26.

Smith, B. D. 1995. The Emergence of Agriculture. New York: Scientific American Library.

Smith, N., ed. 2000. Seeds of Opportunity: An Assessment of the Benefits, Safety, and Oversight of Plant Genomics and Agricultural Biotechnology. Committee Print 106-B. Washington, D.C.: Subcommittee on Basic Research, Committee on Science, U.S. Congress.

Society of Toxicology ad hoc Working Group. 2003. Society of Toxicology position paper: the safety of genetically modified foods produced through biotechnology. Toxicological Sciences 71:2-8.

Solomon E. B., S. Yaron, and K. R. Matthews. 2002. Transmission of *Escherichia coli* O157:H7 from contaminated manure and irrigation water to lettuce plant tissue and its subsequent internalization. Applied and Environmental Microbiology 68:397-400.

Stanley-Horn, D. E., G. P. Dively, R. L. Hellmich, H. R. Mattila, M. K. Sears, R. Rose, L. C. H. Jesse, J. E. Losey, J. J. Obrycki, and L. Lewis. 2001. Assessing the impact of Cry1Ab-expressing corn pollen on monarch butterfly larvae in field studies. Proceedings of the National Academy of Sciences USA 98:11931-36.

Tabashnik, B. E., R. T. Roush, E. D. Earle, and A. M. Shelton. 2000. Resistance to Bt toxins. Science 287:41.

Tabashnik, B. E., Y. Carriere, T. J. Dennehy, S. Morin, M. S. Sisterson, R. T. Roush, A. M. Shelton, and J. Z. Zhao. 2003. Insect resistance to transgenic Bt crops: lessons from the laboratory and field. Journal of Economic Entomology 96:1031-38.

Taylor, S. L. 2003. Safety assessment of foods produced through agricultural biotechnology. Nutrition Reviews 61:(II)S135-40.

Tennant, P., G. Fermin, M. Fitch, R. Manshardt, J. Slightom, and D. Gonsalves. 2001. Papaya ringspot resistance of transgenic Rainbow and SunUp is affected by gene dosage, plant development, and coat protein homology. European Journal of Plant Pathology 107:645-53.

Thrupp, L. A. 2002. Fruits of Progress: Growing Sustainable Farming and Food Systems. World Resources Institute.

Tilman, D. 1998. The greening of the Green Revolution. Nature 396:211-12.

Toenniessen, G. 2000. Vitamin A deficiency and Golden Rice: the role of the Rockefeller Foundation. New York: Rockefeller Foundation.

Toenniessen, G. 2002. Crop genetic improvement for enhanced human nutrition. Proceedings of the XX International Vitamin A Consultative Group meeting, February 2001. Journal of Nutrition 132:2943S-46S.

Toenniessen, G., J. C. O'Toole, and J. DeVries. 2003. Advances in plant biotechnology and its adoption in developing countries. Current Opinions in Plant Biology 6:191-98.

Torrey, J. G. 1985. The development of plant biotechnology. American Scientist 73:354-63.

Tudge, C. 1995. The Engineer in the Garden: Genes and Genetics from the Idea of Heredity to the Creation of Life. New York: Hill and Wang.

Tudge, C. 2000. In Mendel's Footnotes: An Introduction to the Science and Technologies of Genes and Genetics from the Nineteenth Century to the Twenty-Second. London: Jonathan Cape.

Tudge, C. 1998. Neanderthals, Bandits, and Farmers: How Agriculture Really Began. New Haven and London: Yale University Press.

Turney, J. 1998. Frankenstein's Footsteps: Science, Genetics, and Popular Culture. New York and London: Yale University Press.

U.S. Environmental Protection Agency. 2001. *Bacillus thuringiensis* (Bt) Plant-Incorporated Protectants. Biopesticides Registration Action Document, 15 October 2001. Washington, D.C.: U.S. Environmental Protection Agency.

U.S. Food and Drug Administration. 1992. Foods derived from new plant varieties. Statement of Policy. US Federal Register 57:22984-3005.

Valanzuolo, S., S. Catello, M. Colombo, M. Dani, M. M. Monti, L. Uncini, P. Petrone, and P. Spigno. 1994. Cucumber mosaic virus resistance in transgenic San Marzano tomatoes. Fifth International Symposium on the Processing Tomato. ISHS Acta Horticulturae 376:52-ff.

Venter, J. C., M. D. Adams, E. W. Myers, P. W. Li, R. J. Mural, G. G. Sutton, H. O. Smith, M. Yandell, C. A. Evans, R. A. Holt, et al. 2001. The sequence of the human genome. Science 291:1304-51.

Visser, M. 1986. Much Depends on Dinner: The Extraordinary History and Mythology, Allure and Obsessions, Perils and Taboos of an Ordinary Meal. New York: Grove Press.

Wachtel, M. R., L. C. Whitehand, and R. E. Mandrell. 2002. Association of *Escherichia coli* O157:H7 with preharvest leaf lettuce upon exposure to contaminated irrigation water. Journal of Food Protection 65:18-25.

Wang, R. L., A. Stec, J. Hey, L. Lukens, and J. Doebley. 1999. The limits of selection during maize domestication. Nature 398:236-39.

Weaver, W. W. 2000. 100 Vegetables and Where They Came From. Chapel Hill: Algonquin.

White, S., and J. Doebley. 1998. Of genes and genomes and the origin of maize. Trends in Genetics 14:327-32.

Wilkinson, M. J., I. J. Davenport, Y. M. Charters, A. E. Jones, J. Allainguillaume, H. T. Butler, D. C. Mason, and A. F. Raybould. 2000. A direct regional scale estimate of transgene movement from genetically modified oilseed rape to its wild progenitors. Molecular Ecology 9:983-91.

Wilkinson, M., L. J. Elliott, J. Allainguillaume, M. W. Shaw, C. Norris, R. Welters, M. Alexander, J. Sweet, and D. C. Mason. 2003. Hybridization between *Brassica napus* and *B. rapa* on a national scale in the U.K. Science 302:457-59.

Winicov, I. 1998. New molecular approaches to improving salt tolerance in crop plants. Annals of Botany 82:703710.

Woese, K., D. Lange, C. Boess, and K. W. Bogl. 1997. A comparison of organically and conventionally grown foods—results of a review of the relevant literature. Journal of the Science of Food and Agriculture 74:281-93.

Wood, D. W., J. C. Setubal, R. Kaul, D. E. Monks, J. P. Kitajima, V. K. Okura, Y. Zhou, L. Chen, G. E. Wood, N. F. Almeida Jr., L. Woo, Y. Chen, I. T. Paulsen, J. A. Eisen, P. D. Karp, et al. 2001. The genome of the natural genetic engineer *Agrobacterium tumefaciens* C58. Science 294:2317-23.

Ye, X., S. Al-Babili, A. Klöti, J. Zhang, P. Lucca, P. Beyer, and I. Potrykus. 2000. Engineering the provitamin A beta-carotene biosynthetic pathway into carotenoid-free rice endosperm. Science 287:303-5.

Yeh, S., B. A. Moffatt, M. Griffith, F. Xiong, D. S. Yang, S. B. Wiseman, F. Sarhan, J. Danyluk, Y. Q. Xue, C. L. Hew, A. Doherty-Kirby, and G. Lajoie. 2000. Chitinase genes responsive to cold encode antifreeze proteins in winter cereals. Plant Physiology 124:1251-64.

Zambryski, P., M. Holsters, K. Kruger, A. Depicker, J. Schell, M. Van Montagu, H. M. Goodman. 1980. Tumor DNA structure in plant cells transformed by *A. tumefaciens*. Science 209:1385-91.

Zambryski, P., H. Joos, C. Genetello, J. Leemans, M. Van Montagu, and J. Schell. 1983. Ti plasmid vector for the introduction of DNA into plant cells without alteration of their normal regeneration capacity. EMBO Journal 2:2143-50.

Zohary, D., and M. Hopf. 1993. Domestication of Plants in the Old World: The Origin and Spread of Cultivated Plants in West Asia, Europe, and the Nile. Oxford: Clarendon Press.

Zuckerman, L. 1998. The Potato: How the Humble Spud Rescued the Western World. New York: North Point Press.

ABOUT THE AUTHORS

Nina Fedoroff is a leading geneticist and molecular biologist who has contributed to the development of the techniques used to study and modify plants today. She received her Ph.D. from the Rockefeller University, and, as a postdoctoral fellow at the Carnegie Institution of Washington, she successfully sequenced one of the first animal genes. Switching to plants, she investigated the "jumping genes" discovered in corn plants by geneticist Barbara McClintock in the 1940s. She isolated the DNA of these mobile genes and then continued to study their structure, their movement, and how they are controlled. In 1995 she joined the faculty of the Pennsylvania State University, where she focused on genes that protect plants from biological and nonbiological stresses. Fedoroff is a member of the National Academy of Sciences and is currently serving on the National Science Board.

Nancy Marie Brown was trained as a medievalist but has worked since 1981 as a science writer. Until recently she was editor of *Research/ Penn State* magazine, for which she collaborated with Nina Fedoroff on a series of articles on genetics. Her first book, *A Good Horse Has No Color: Searching Iceland for the Perfect Horse*, was published in 2001. She is currently editing the memoirs of an herbalist, working on a book about the science of horse breeding, and writing about science and nature for children. She lives in a restored farmhouse on 100 acres in Vermont.